ADVERSE REACTIONS

ADVERSE REACTIONS

THE FENOTEROL STORY

NEIL PEARCE

AUCKLAND UNIVERSITY PRESS

First published 2007

Auckland University Press
University of Auckland
Private Bag 92019
Auckland
New Zealand
www.auckland.ac.nz/aup

ISBN 978 1 86940 374 4

National Library of New Zealand Cataloguing-in-Publication Data
Pearce, Neil.
Adverse reactions : the Fenoterol story / Neil Pearce.
ISBN 978-1-86940-374-4
1. Fenoterol. 2. Asthma—Research. 3. Asthma—Research—New
Zealand. 4. Asthma—Mortality. 5. Asthma—New Zealand—
Mortality.
I. Title.
616.238061—DC 22

Cover design: Spencer Levine, Base Two

Printed by Printlink Ltd, Wellington

*This book is dedicated to my parents
(Arthur and Peggy Pearce)
and my family (Lynette, Anna, Lucy)*

Contents

List of Figures

List of Tables

ADEC   Australian Drug Evaluation Committee
BMJ   *British Medical Journal*
CRM   Committee for Review of Medicines
ERI   Epidemiology Resources Inc
FDA   Food and Drug Administration
GP   General Practitioner
HRC   Health Research Council
IARC   International Agency for Research on Cancer
IEA   International Epidemiology Association
ISPE   International Society for Pharmacoepidemiology
JAMA   *Journal of the American Medical Association*
MARC   Medicines Adverse Reactions Committee
MPS   Medical Protection Society
MRC   Medical Research Council
NEAC   National Ethics Advisory Committee
NEJM   *New England Journal of Medicine*
NZMJ   *New Zealand Medical Journal*
PHA   Public Health Association
WHO   World Health Organisation

Beta agonists (and brand-names)

Fenoterol (Berotec)
Isoprenaline/isoproterenol (Medihaler/Medihaler Iso)
Orciprenaline/metaproterenol (Alupent)
Terbutaline (Bricanyl)
Salbutamol/albuterol (Ventolin/Proventil)

In early 1989 in my first few weeks as Minister of Health, officials advised me that New Zealand research was soon to be published claiming safety concerns over a widely used asthma drug called fenoterol.

The evidence produced by Neil Pearce and others claimed to show a link between the use of fenoterol and the death rate among asthma sufferers in New Zealand. That death rate was much higher than that in comparable countries at the time. The work of Neil Pearce and his colleagues seemed to show that this 'epidemic' of deaths had coincided almost exactly with the widespread use of fenoterol in New Zealand, starting in 1976.

The significance of the evidence presented by this group of investigators was obvious. Their initial research set in train a series of reviews which enabled me to announce ten months later that fenoterol was to be removed from the Drug Tariff. This greatly reduced the availability of the drug and was followed very quickly by a dramatic reduction in the death rate.

This important decision, and the subsequent fall in the death rate, was a direct outcome of the work of this group of investigators operating under very difficult circumstances. This book is the story of a remarkable saga of medical detective work as told by one of the lead players, Neil Pearce.

Asthma is a debilitating disease which affects thousands of New Zealanders, and a heavy focus has been rightly placed on developing effective treatment for it. Thus, drugs have been developed to help asthma sufferers cope with the disease and free them from its burdens. But there have been instances - as in the case of fenoterol – where there have been real doubts about the efficacy of a drug, and indeed claims that it raised, rather than reduced, the chances of dying from asthma.

The lessons we can take from the fenoterol story are as relevant today as they were when the issue was in the news headlines seventeen years ago.

Neil Pearce's book raises issues of drug safety - not only here in New Zealand, but also internationally. It gives an insight into the big-money world of drug development and marketing, where the stakes are high and pharmaceutical companies strive to develop new treatments in competition with each other.

It also reinforces the significant role which science has to play not just in developing new treatments but in assessing them after their release. There is likely to be a natural tension between the cautions of science and the ambitions of the drug industry. But this is only healthy, and it is essential for both the continuous improvement in drug treatments and the protection of public safety.

In a climate where the next 'wonder drug' is talked up in the media and advertised to the public, often without its potential risks being fully known or widely discussed, this book is a reminder not only to be vigilant and ensure that new drugs are subjected to the most rigorous examination before they are made available to the public, but also to continue that scrutiny long after they have been released from the laboratory and the clinical trials to assess how they work in the real world of medical practice.

The fenoterol story is one which needs to be told in full, and its lessons need to be heeded. I congratulate Neil Pearce on this book.

Helen Clark
Prime Minister

Acknowledgements

This book has been a long time in the making and there are many people I should acknowledge for their help with it, or with the events described in it.

The directors of the Wellington Asthma Research Group were Richard Beasley, Carl Burgess, Julian Crane and myself. The group was formed in 1988 specifically to do the studies of fenoterol and asthma deaths. After the events described in this book, which took place between 1988 and 1990, it went on to do other types of asthma research until I moved to another university in 2000.

Rod Jackson also played a crucial role in the events described in this book. He is a rare scientist with integrity who was prepared to change his mind because of new evidence. Michael Hensley played a similar role in Australia, despite strong criticism from his colleagues there. Others also played important supportive roles, including Eru Pomare, Robert Beaglehole, Richard Doll, Paul Stolley, Allan Smith, Ken Newell and Peter Davis.*

Al Morrison interviewed Paul Stolley when he visited Wellington in 1989, and I am grateful for permission to quote from that interview. In 1991, Elizabeth Heseltine interviewed many of the other key players, and all of the other quoted interviews were done by her.

I would also like to thank: Elizabeth Caffin and the team at Auckland University Press; Andrew Mason for his major editing job; Brian Easton, Geoff Fougere, Franco Merletti and Lorenzo Richiardi for their comments on the draft manuscript; and PHARMAC, the New Zealand Pharmaceutical Management Agency, for information on fenoterol sales.

---

* Peter Davis asked me to write a chapter about the fenoterol story for a book, *For health or profit: The pharmaceutical industry in New Zealand*, which was first published in 1992, and then in modified form in 1996, as *Contested ground: public purpose and private interest in the regulation of prescription drugs*. Some of the material from these book chapters appears, substantially modified and expanded, in this book.

Between 1979 and 2000 I was employed at Otago University's Wellington School of Medicine. Since 2000 I have been at Massey University, which has given great support for my research; this in part allowed me to take the time to finish this book. During most of this period my salary has been funded by various grants from the Health Research Council of New Zealand; my sabbatical in Lyon in 1993 was supported by a Visiting Scientist Award from the International Agency for Research on Cancer, and my sabbatical in Turin in 2005 was supported by the Progetto Lagrange, Fondazione CRT/ISI.

Finally, and most importantly, I want to thank my wife, Lynette Shum, and our daughter Anna Shum Pearce, who was about six at the time that the fenoterol saga occurred. Lynette and Anna are barely mentioned in the book, but they were there the whole time and I couldn't have done it without them. Our second daughter, Lucy Shum Pearce, came later, but has had to endure years of me telling the stories that are in this book. My father, Arthur Pearce (1903–1990), died in the middle of the fenoterol story, and never lived to see that we were proved right. My mother, Peggy Pearce (born 1916), is still alive today. I want to thank them all for their love and support over many years.

FOOLS RUSH IN
I've been to Dublin only once, but it was a classic Irish tourist experience. I spent the first night drinking beer in Temple Bar, surrounded by tour groups, but the next night I managed to find the real thing. The corner pub near the backpackers I was staying in was crammed with locals and full of Irish music. As the band sank into the third verse of 'Whisky in the jar', and I sank into my third pint of Guinness, I began to think about how strange the previous couple of years had been. I couldn't think about this for long, as I was cornered by a visitor from County Mayo, 'where you stand on the cliffs and the wind whistles in from the Atlantic and makes you feel small'. He bought me another five pints of Guinness while he told me his life story. The climax was that he had come to Dublin that night, because the following morning he was going to have to put his mother into psychiatric care. His brother and his sister didn't have the courage to do it, so he had to. But that was tomorrow, and tonight he could get drunk. He seemed too Irish to be true – maybe he was working for the Irish Tourist Board – but he was real, and he did not know what had hit him. 'Why me?' I knew how he felt.

The story began for me at about 10.30 am on Friday, 8 April 1988. I was sitting in my office in the Department of Public Health at the Medical School in Wellington when someone arrived unexpectedly and started talking about a drug called fenoterol which might be causing an epidemic of asthma deaths in New Zealand. I knew nothing about asthma – I usually got it confused with eczema – but I had heard about the epidemic of deaths. There had been a lot of publicity about it during the 1980s, and a friend of mine had done some work trying to find out the cause. It sounded an interesting problem, so I agreed to help out.

Two years later, fenoterol had been restricted in New Zealand and the death rate had fallen by more than half. The drug was also restricted in Australia and Japan, and the company which makes the drug had agreed to halve its dose in other countries. The United States Food and Drug Administration had held hearings into the safety of beta agonists (the class of asthma drugs that fenoterol belongs to). Editorials had appeared in some of the major international medical journals calling for the safety of drugs like fenoterol to be reassessed.[1]

This book is the story of those two years. Several journalists who got involved in the story suggested that it would make a good book, but they were all too busy to do it. In 1993, while on sabbatical at the International Agency for Research on Cancer in Lyon, France, I decided to write a book myself, or at least to document the story before I started to forget it all. What I wrote was too serious, too technical – and too angry. It needed more distance. So I sat on it for twelve years, and revised it again when I was on sabbatical at the University of Turin in Italy in 2005.

It is a story which raises many questions about the role of drug companies, doctors, researchers, the government and the public in issues of drug safety. It has a long history, and in the next few chapters I'll go over the history before getting back to my small part in it. It is a New Zealand story, but it is also about the problems with drug safety internationally, and about the contest between money and science in medical research. For me it is also very much a personal story. I'm not claiming to tell the whole story, just the part of it that happened to me.

CHAPTER 1 Some History

If I hadn't believed it, I wouldn't have seen it. – YOGI BERRA[1]

The word 'asthma' comes from a Greek word meaning 'panting'. Asthma has been around for thousands of years, but we still don't know much about what causes it, or what makes it better or worse. Doctors can't even agree on how to define asthma. Salter, a 19th-century English doctor, wrote:

> The circumstances under which asthma may occur are so various and the features of different causes so peculiar, and impart to those cases such an individuality, that all writers are tempted with more or less success to make some classification of its different varieties . . . It seems to me a subject in which authors have done more in the way of reading each other's books than in scrutinizing their own patients.[2]

Nowadays, asthma is regarded as a disease of the airways (the bronchial tubes in the lungs) which involves difficult breathing that gets better or worse from time to time. So the key feature of asthma is that it 'comes and goes'. This makes it different from other lung diseases, such as chronic bronchitis or emphysema, which involve (more or less) permanent breathing problems. Asthma often involves 'wheezing'. This is a high-pitched whistling sound heard during breathing, especially when breathing out. In people with asthma, the airways are 'twitchy' or irritable, and they react to some 'triggers' by narrowing (called bronchospasm), the tissue around the bronchial tubes becomes inflamed, and mucus blocks the airways. These three problems – bronchospasm, inflammation and mucus – are the key features of asthma.[3]

Aretaeus of Cappadocia, a Greek physician in Rome in the second century AD, provided the first accurate written description of an asthma attack:

> If from running and exercise, and labour of any kind, a difficulty of breathing follows, it is termed asthma, and the disease orthopnoea, itself is likewise called asthma . . . The precursory symptoms of this disease are weight at the chest, an unwillingness to attend to one's ordinary vocation, or to business altogether, an uneasiness of respiration in running, or going uphill . . . Under increasing disorder the cheeks flush, the eyes are prominent as in cases of strangulation, a snoring is heard while they are awake and the evil is much augmented during the sleep. The voice indicates the presence of mucus, is feeble and indistinct . . . when the disease takes a favourable turn the cough is longer, though less frequent, with an excretion of humid matter in great quantity . . . the breathing becomes rare, gentle, but there is asperity of the voice.[4]

An asthma attack can be triggered by many different things: pollens, grasses, moulds, animal dandruff, urine, hair or feathers, house dust mites, exercise, changes of temperature, air pollution, colds, dusts, fumes, tobacco smoke and some foods like peanuts. It is still not known exactly why this happens, or how the triggering of an asthma attack can be prevented.

ASTHMA TREATMENT

Until recently, the treatment of asthma has been based mainly on guesswork, and many different treatments have been tried. For example, Caelius Aurelianus, a 5th-century physician, advised using 'simple clyster, pultices, laxatives, cupping of the chest with scarifications as well as steam, fomentation, sponging, rubbing of the arms and during paroxysms, where strength permits, venesection'.[5]

One of the earliest known examples of successful asthma treatment was in the 16th century, when Gerolamo Cardano was called from Italy to Edinburgh to treat John Hamilton, the Archbishop of St Andrews. His treatment included diet, purging, regular exercise, sleep and substitution of unspun silk for feathers in the mattress. This kept him alive long enough to be hanged by the Scottish Reformers.[6]

In the last few centuries, treatments for asthma have included 'foxes lungs, Syrup of Garlick, Tincture of Lavender, Saffron Lozenges and Smoked Amber with Tobacco' (17th century), 'bloodletting and gentle vomits' (18th century) and 'whiffs of chloroform, lobelia and morphine' (19th century).[7] Other drugs that were tried in the 19th century included marijuana, petroleum, oil and various 'asthma powders' made from extracts of plants such as datura. These were sold without restriction in shops and on street corners. Most of them were never properly tested and probably did more harm than good. I have a tin of Neil's Asthma Powder which must be more than 50 years old. It contains powdered lobelia and stramonium, is labelled as a poison and has instructions to take half to one teaspoon:

> . . . put on some red embers or burnt in any way, inhale smoke through nose and mouth . . . The amazing efficacy of this remedy is due to the inclusion of a secret and little used herb in conjunction with a variety of specially chosen herbs that have been famed down the centuries for their healing and penetrating qualities.

Until modern treatment was introduced in the 1940s, asthma was a crippling disease which could make it impossible to go to school, get a job or have a life. A severe asthma attack could be frightening, leaving the sufferer gasping for breath and seemingly close to death. However, asthma hardly ever killed anyone, and most people with asthma lived into old age. Most studies of asthma deaths involve the 5- to 34-year age-group because the diagnosis of asthma death is very accurate in this age-group, but it is much less accurate in young children and those over 35 years.[8] As Figure 1 (overleaf) shows, in the first half of the 20th century, in people aged 5-34 years, there was less than one asthma death per 100,000 people per year in the United Kingdom and in New Zealand.[9]

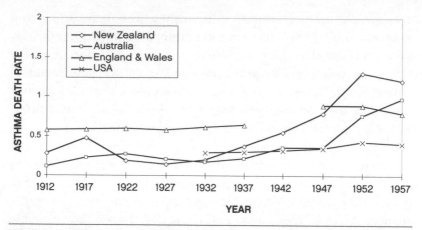

FIGURE 1: Asthma deaths per 100,000 per year in the 5- to 34-year age-group in New Zealand, Australia, England and Wales, and the USA between 1910 and 1960 (there is a gap in the data for England and Wales because of the unreliability of statistics during the Second World War)

THE BETA AGONISTS

For most of the 20th century it was believed that bronchospasm was the main problem in asthma. In some ways this is obvious, because bronchospasm is what happens when someone has an asthma attack. The muscles in the airways tighten up, which makes them narrower, and it is more difficult for the air to get through. In a very severe attack, some airways tighten up so much, and get so full of mucus, that no air can get through. So if there was a drug that would relax the muscles in the airways, allowing more air to get through, then the problem of bronchospasm could be fixed. This is what beta agonist drugs do. They belong to a wider group of drugs known as 'bronchodilators', which all have this characteristic. They are also known as 'relievers', because they relieve the symptoms of asthma (i.e. bronchospasm), and in the United States as 'sympathomimetics'. The beta agonists do not fix the problems of mucus and inflammation – they can even make these problems worse – but they do relieve the bronchospasm and make the asthma sufferer feel better and breath more easily.

Nowadays the most commonly used beta agonist is 'salbutamol', but this came on the market only in about 1969.[10] The first

6

beta agonists to be marketed, in the 1940s, were 'adrenaline' and 'isoprenaline'. These are generic names which are used world-wide – except that the generic names are often different in the United States, where isoprenaline was known as 'isoproterenol'. However, the drugs are sold under various brand-names in different countries. For example, isoprenaline was sold as 'Medihaler' and 'Medihaler Iso'. Salbutamol is mostly marketed under the name of 'Ventolin'.*

Isoprenaline was introduced into the United Kingdom as a spray in a 'squeeze bottle' in 1948, and became available in New Zealand soon after. It was good at relieving symptoms and making asthma more tolerable, and there was a rapid increase in sales in the 1950s. Very soon, concerns were raised about the safety of isoprenaline, and the asthma death rate increased a little during the 1950s, as Figure 1 shows.[11] However, the increase in asthma deaths went unnoticed, because it was fairly small. There was great enthusiasm for this new form of treatment which seemed to be so effective. Not only did the patients like it, the doctors did too. Most of the time, being a doctor, and especially a general practitioner, is a thankless task, because most patients have problems that can't be fixed. But here was a drug which seemed actually to work; it made the patients feel better, and for many of them it made the difference between having to stay in bed and being able to get up and have a life.

Enthusiasm for beta agonists grew when the pressurised metered dose inhaler was introduced in the early 1960s. This is also known as an 'aerosol', 'inhaler' or 'puffer' in many countries, and as a 'nebuliser' in the United States – in other countries

---

* The brand-name is just a marketing tool to try to make the drug sound more attractive to doctors and patients. The use of different brand-names in different countries can cause great confusion. When the link between thalidomide and birth defects was established in the early 1960s, for example, some pregnant women kept on taking the drug because it was sold under different brand-names in different countries and they did not realise that they were taking thalidomide (which is the generic name). The system of brand-names continues today because drug companies believe that it helps to increase their sales by distinguishing their version of a particular drug from the versions produced by other companies.

'nebuliser' refers to an electrically driven 'puffer' for delivering high doses in a severe attack. This was a much easier way to take the drug, and sales soared. A new beta agonist known as 'orciprenaline' was also introduced around this time (its generic name was 'metaproterenol' in the US, and it was sold under the brandname of 'Alupent'), but isoprenaline continued to account for most of the beta agonist sales.

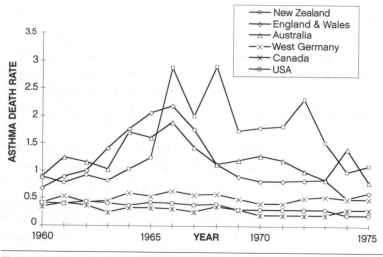

Figure 2: Asthma deaths per 100,000 per year in the 5- to 34-year age-group in New Zealand, England and Wales, Australia, West Germany, Canada and the United States during the 1960s and 1970s

THE 1960S EPIDEMICS

Suddenly, in the 1960s, there were epidemics of asthma deaths in six countries: New Zealand, England and Wales (which are counted as one country in UK health statistics), Australia, Scotland, Ireland and Norway. The data for three of these countries, and for three countries that did not have epidemics (West Germany, Canada and the US), are shown in Figure 2. These epidemics occurred very suddenly, and were completely unexpected, because the death rate from asthma had been so low and so stable for nearly a century. There had been the small increase in deaths during the 1940s and 1950s, but an epidemic of asthma deaths had never happened before.

One of the first published reports of the epidemic came in March 1967 in a letter to the *British Medical Journal* from two British researchers, who warned: 'We suspect that patients with asthma may be killing themselves by the excessive use of sympathomimetic agents in the form of metered or pressurized aerosols containing isoprenaline, orciprenaline, or adrenaline'. They described four patients who had died and eight who had been admitted to hospital with severe asthma attacks. They argued:

> A label should be put on each inhaler . . . specifying that not more than one puff should be taken and at not more than half-hourly intervals . . . so far as we are aware there are no such warnings given on any of the aerosol containers or with the instructions, with one or two exceptions, where the warning is not very prominent.[12]

This raised enough interest for Sir Richard Doll to get involved. Doll was the most famous epidemiologist of his generation (he died in 2005). He should have won the Nobel Prize, but didn't, because epidemiology is not highly valued by the medical establishment.

The word 'epidemiology' comes from the word 'epidemic', and in the 19th century it was used for the study of the causes of epidemics of cholera, typhoid and other 'diseases of poverty'. However, epidemiology has grown in popularity and importance since the 1950s when Sir Richard Doll and others used it for the study of the causes of non-communicable diseases like heart disease, cancer and asthma.

Epidemiologists study the causes of diseases by studying populations and comparing the risk of getting a disease (or the risk of dying) in different population groups. For example, the link between smoking and lung cancer[13] was established in studies which compared the risk of developing lung cancer in smokers with the risk in non-smokers. Of course, the ideal way to do this would have been to take 100,000 people, toss a coin 100,000 times and persuade all of the 'heads' to smoke and all of the 'tails' not to, then follow them for 40 years and see what happened.

However, a randomised trial like this would be impossible – and unethical. So, instead, a population survey is necessary to find out who smokes and who doesn't. The epidemiologist then follows the population over time to find out who develops lung cancer. Doll did this with a famous study of British doctors and found that those who smoked were ten times more likely to get lung cancer than those who didn't smoke.[14]

The problem, of course, is that people who smoke are different from people who do not. Smoking has not been randomised by tossing a coin – it is something that people have chosen to do. So there might be something else that is different about smokers which explains why they have a higher risk of lung cancer. However, it is usually possible to check this out. Information on other causes of lung cancer (such as asbestos exposure) can be collected and adjusted for in the data analysis. Even if information is not obtainable on all of the known causes of lung cancer, estimates can be made on how strong a 'bias' may have occurred, and whether this is strong enough to explain the study findings.

STUDIES OF THE EPIDEMICS

Epidemiology was the only realistic approach for studying the causes of the 1960s epidemics of asthma deaths, since it would have been impossible and unethical to do a randomised trial, even if the likely cause of the epidemics had already been known. Frank Speizer and Richard Doll, at Oxford University, studied the time trends of deaths in the 5- to 34-year age-group,[15] and concluded that the epidemic was real and was not due to changes in diagnostic criteria (the way that doctors diagnose asthma), certification (the way that deaths are recorded) or coding practices (the way that deaths are classified). They also concluded that it was unlikely to be due to a sudden increase in asthma prevalence, since there was no evidence that any increase had occurred. Rather, the evidence suggested that the epidemic of deaths was due to an increase in case-fatality (the percentage of people with asthma who die) caused by new methods of treatment.

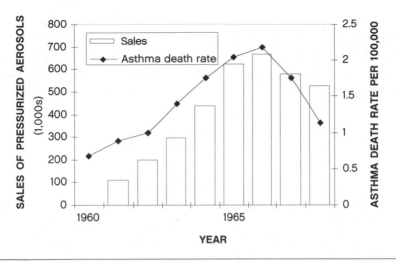

Figure 3: Sales and prescriptions of beta agonist inhalers (thousands) and asthma deaths per 100,000 per year in the 5- to 34-year age-group in England and Wales during the 1960s

In particular, they found that the epidemic started after the beta agonist inhalers were introduced in 1961, and the rise in asthma deaths paralleled the increase in sales (see Figure 3). They also noted that the increase in deaths was greatest in the 10- to 19-year age-group and that 'at these ages children have begun to act independently and may be particularly prone to misuse a self-administered form of treatment'.

A particular concern was that the inhalers were available 'over the counter' in many countries, including the United Kingdom. They could be bought from the pharmacy without a doctor's prescription. This encouraged patients to keep using their inhalers, and to buy another one at the pharmacy, rather than to get help from a doctor when their asthma got worse.

The new inhalers were very likely to be overused, because of both the way that the beta agonists worked and the way they had been packaged. The beta agonists provide almost 'instant relief' to someone suffering an asthma attack, and it is easier and cheaper to 'take another puff' than go to the doctor. This was particularly true in New Zealand, because (until recently) the medicines were free, but there was a charge for seeing the family doctor. It is very unusual for someone with asthma to swallow a whole bottle of pills – before this happens they will usually realise that something is seriously wrong and see a doctor. But it is quite common for someone having an asthma attack to use up an entire inhaler, rather than go to the doctor. For example, a study in Wellington Hospital in 1989 found that 22 per cent of the people coming to hospital with a severe asthma attack had taken more than 60 puffs in the previous 24 hours before seeking help, when the recommended dose was only two puffs four times per day.[16]

In response to these concerns about beta agonists, the British Committee on Safety of Drugs sent a pamphlet to all doctors in the United Kingdom in June 1967 warning of the possible dangers of overuse of inhalers, and these were removed from over-the-counter sale in December 1968. The death rate fell following these moves (see Figure 3).

In a subsequent study, Inman and Adelstein found that 87 per cent of the total sales of pressurised aerosols in England and Wales were of products containing isoprenaline, whereas 8 per cent contained orciprenaline, and 4 per cent contained adrenaline. There appeared to be a close parallel between over-the-counter sales of inhalers and the epidemic of deaths (see Figure 3). They concluded that 'the excess deaths . . . were likely to have been the result of overuse of pressurized aerosols and that the subsequent decline in mortality has resulted from a greater awareness by doctors and patients of the dangers of overuse'.[17]

Further support was given to this hypothesis, which might be called the *beta agonist hypothesis*, by a British study of 174 asthma deaths in young people which found that 84 per cent of the people who died had been using beta agonist inhalers shortly before they died, and that there was some evidence that the inhalers had been used excessively.[18] Another British study in 1968–69, when the epidemic had already begun to decline, suggested that overuse of beta agonist inhalers had contributed to 37 per cent of deaths in those aged 5–34 years.[19]

So why was this happening? The patients liked their inhalers and their doctors noticed that their asthma improved when they used them. Most doctors would say that they had never noticed a problem with their patients taking isoprenaline. So if the drug was so good, and no one had seen a problem, did it really matter that some patients overused it? Why would the drug cause asthma deaths? There were three possible versions of the *beta agonist hypothesis*, each involving a different potential mechanism by which overuse of beta agonist inhalers could increase the risk of death. These were: the *delay hypothesis*; the *toxicity hypothesis*; and the *severity hypothesis*.

The *delay hypothesis* suggested that, while the beta agonist inhalers might not be directly toxic (i.e. they might not cause death directly), they could be so effective at relieving symptoms (without curing the disease) that people might delay seeking medical help until it was too late. Most people who die from asthma die on the way to hospital, or when they are waiting for the ambulance, and in many cases they might have lived if they had sought help half an hour earlier. The delay hypothesis is sometimes used to argue that the epidemics were not due to overuse of asthma drugs; rather, the problem was with the patients who delayed seeking help when they needed it. However, if overuse of the drug had caused such delays, then the drug had indirectly caused the deaths. So this was one possible explanation of why the introduction of the beta agonist inhalers could have caused the epidemics.

Most respiratory physicians were prepared to accept the delay hypothesis, although they blamed the patients for the delays rather than the drugs they had prescribed for them. However, most were not prepared to believe that beta agonists were directly toxic (the *toxicity hypothesis*). Most deaths during the epidemic occurred in patients with very severe asthma, and it was assumed that the deaths were due to asthma itself rather than to drug toxicity. Furthermore, if someone is having an asthma attack, and makes it to hospital, they are usually given very high doses of beta agonists in the emergency department, but asthma deaths in hospital are very rare.[20]

One explanation for this apparent anomaly was provided by British researchers[21] who showed that it was possible to give large doses of isoprenaline to dogs breathing normal air, but that much smaller doses caused fatal side-effects on the heart if the dogs were made hypoxic (short of oxygen). When the dogs were autopsied, the damage to the heart was seen to be similar to that described in some children dying of asthma during the epidemic. Hypoxia occurs in severe asthma attacks because of the difficulty of passing air through the narrowed and inflamed airways. It is not known exactly why isoprenaline should cause death under conditions of hypoxia, but not in more normal circumstances. However, what may be happening is that the side-effects of isoprenaline result in an increased workload on the heart.[22] These are the beta-1 effects of beta agonists; they are different from the beta-2 effects, which relax the airways and let the patient breathe more easily. A further problem is that isoprenaline can make hypoxia worse. If the heart is under a lot of strain from the side-effects of the drug, and at the same time it is deprived of oxygen, then it may not be able to cope with the increased workload, and heart failure may result.

These studies suggested why most deaths occurred outside hospital, under conditions of hypoxia (i.e. when people were having an asthma attack and were short of oxygen), whereas large doses of beta agonists in hospital might be safe if oxygen was given at the same time. They also suggested that acute toxicity (or overdose) was unlikely to occur, unless there had also been delays causing the asthma to deteriorate and hypoxia to occur. So isoprenaline was dangerous only in patients who already had very severe asthma, and who were likely to have bad attacks which made them hypoxic. It was a very safe drug if a person did not have asthma.

This point has been very difficult for many doctors to understand and to accept. If someone is having a bad asthma attack and they die, it is reasonable to conclude that they died of asthma. However, in many cases the patient might not have died, or might have survived long enough to make it to hospital, where oxygen was available, if they had not been overusing isoprenaline. In this situation, death was caused by a combination of a bad asthma attack and overuse of isoprenaline.

The third possible explanation of why overuse of beta agonist inhalers might cause asthma deaths was the *severity hypothesis*. This suggested that use of beta agonist inhalers could make asthma worse in the long term. For example, one study found that overuse of isoprenaline could make asthma more severe.[23] The reasons for this are not clear, but it is possible that isoprenaline, by opening the airways, could increase exposure to factors which trigger off asthma, or may increase the sensitivity to these factors.

These problems have been described by Paul Stolley, an American pharmacoepidemiologist (an epidemiologist specialising in drug side-effects):

> I have a simple model in my mind of beta agonists and it has to do with an experience most of us have using nasal spray when we have a cold . . . after three or four days, you find you're using it more frequently to unclog your nose, and then eventually your nose is completely clogged a half-hour after you use it. You get a rebound phenomenon . . . people become addicted to them and are completely obstructed. We admit them to hospital on steroids and wean them off these nasal sprays, but even if you don't get addicted, most people recognise themselves that they're getting worse from using them after a few days.
>
> There may be a phenomenon like that going on, in some asthmatics, who have to use more and more to get less and less relief. And then they get addicted to [the inhalers], and they use them all the time . . . I don't mean . . . habitually addicted, but the fact that once they try to withdraw they may have problems.[24]

INITIAL CONTROVERSIES

Despite the epidemiological evidence, and the existence of several possible explanations for the mechanisms, the hypothesis that beta agonists were causing epidemics of asthma deaths was regarded with scepticism by most doctors. They simply did not believe it. As Paul Stolley remarked: 'It's very hard if you are prescribing these drugs, and you see the relief that the patient gets, to think that you are doing harm. And it's also hard for the patients to think that.'[25]

In Australia, there was a major controversy about the asthma mortality epidemics, and many leading respiratory physicians denied the possible role of asthma drugs. For example, Read published a letter in the *British Medical Journal* in June 1967 stating: 'I deplore the present "hue and cry" approach to the problem of deaths from asthma, and still more the emotional seizing on one valuable form of therapy.'[26] There was also heated correspondence in the *Medical Journal of Australia*, which ended with Dr Munro Ford concluding in exasperation:

> [Read] will just have to stop talking about asthma as if he is the last word on the whole subject, and be less hypercritical of other workers whose views on management do not happen to coincide with his own . . . we must not let the wounded ego of anybody whose favourite brand of treatment has been questioned . . . stifle suspicions about any drug. [27]

Most doctors did not believe that the epidemics were caused by isoprenaline, but no one was able to think of a plausible alternative explanation. The epidemics occurred in six countries around the world, but not in neighbouring countries, and followed nearly a century of very stable, and very low, asthma death rates. Several leading epidemiologists had examined the data and concluded that the epidemics were real, were not due to changes in prevalence, and were most likely to be due to a change in the case-fatality rate. This in turn seemed most likely to be due to a sudden change in management, such as the introduction of a new drug, and this remained the only plausible explanation. Yet, the scepticism of the doctors was partly justified by anomalies in the epidemiological data itself. In particular, large sales of aerosols had occurred in the United States and several other countries which did not experience an epidemic of deaths.

THE ISOPRENALINE FORTE HYPOTHESIS

This confusing situation was resolved by Paul Stolley when he was on sabbatical at Oxford, working with Sir Richard Doll. As he later recounted:

In my first month there, Richard Doll gave a seminar and he recounted the 1960s epidemic, which he thought was due to the introduction of the isoprenaline inhalers . . .

[When he was finished] he looked at me in the audience and said, 'You know, Paul, your doctors [in the US] must use drugs more carefully because these inhalers went on the market at the same time and you don't have an epidemic'. I said to the audience, 'I've just done a study of drug prescribing in the community, and I can assure you that American doctors as a group don't use drugs any more carefully than any other doctors. If we didn't have the epidemic and we used these inhalers then there must be something wrong with the beta agonist hypothesis.'

Doll said, 'It is interesting, isn't it? Maybe it's something you can work on.'

So that's what I did in my sabbatical. I actually thought that he was wrong and that either the [inhalers] didn't account for the epidemic, or there was something different about the [inhalers] in the US that made them safer . . .

While I was working on this, I was living in a little village on a road from Oxford to Leicester. It had about 200 people in the village, and we were living in an old 15th- or 16th-century house . . . I was getting to know my neighbours a little bit, and they knew I was a doctor.

So they came to my house on a Sunday and asked me to go to a man's cottage. And I got to the man's cottage and he was dead. He was 24 years old. He had been using isoprenaline. He had been in church and his asthma had been acting up. He was inhaling this stuff so many times during the service that in the singing his wife asked him to leave because she thought he was disturbing the other members of the church. He felt ill and went home, and when she came home, he was in the bedroom with the door closed and she didn't disturb him. At about six or seven that evening, she went in. She was afraid he might be dead, and he was dead.

This happened . . . at the time that I was wondering whether I should use my sabbatical this way, because I had three or four other things I wanted to do. I was discussing this with my wife, and I said, 'Well, even for a non-believer there may be a message here.'

The point is, the reality of the epidemic became clear to me in personal terms. I had to go there and say, 'This man has died.' This was not an abstraction. I wasn't six thousand miles away in my study. I was in England and the epidemic was still continuing, and I was convinced that it was an important problem.[28]

So Stolley obtained all the sales data for beta agonists, by writing to all the companies to find out exactly what was in the inhalers:

I was amazed to find that 30 per cent of the English market was this preparation called 'forte', and you couldn't tell from the label that it was five times stronger than the regular strength. So if you had a 9-year-old asthmatic who was wheezing more, and you thought you'd give him something stronger, then instead of writing 'Isoprenaline', you wrote 'Isoprenaline forte'. It was like giving that little boy five [inhalers] all at once. That is almost unheard of in pharmacology, that your next strength is five times the [original] dose.[29]

The high-dose (forte) version of isoprenaline had been licensed in only eight countries.[30] Six of these (England and Wales, Ireland, Scotland, Australia, New Zealand, and Norway) had epidemics of deaths following the introduction of the drug. In the other two (the Netherlands and Belgium), the drug was introduced later than in the other countries, after publicity about the possible role of overuse of beta agonists in the United Kingdom mortality epidemics, and sales were low. No epidemics of deaths occurred in countries such as Canada and the United States in which isoprenaline forte was not licensed.

THE REACTION TO STOLLEY'S WORK

Stolley wrote a paper to report what he had found. It was rejected by the *British Medical Journal* and the *Lancet*, but was finally published in the *American Review of Respiratory Disease* in 1972.[31] As Stolley later recounted:

> The Riker company that made this drug co-operated with me when I first wrote to them, because I said that [the theory that] these [inhalers] were the cause of the epidemic was inconsistent with the American experience. So they thought I was going to destroy [the theory], and they gave me lots of data. Then I sent them the manuscript when I finally finished and they told me they were going to sue me. I had to check with my university (I was at Johns Hopkins School of Public Health at the time), and make sure the lawyers would defend me so I wouldn't go broke.
>
> They didn't sue me, but right after I published it, I got attacked by pulmonary physicians. I presented this at a Thoracic Society meeting in Colorado, and I was attacked mainly for my motives. And then a lot of people gave testimonials to the drug saying [that they had] been using this drug for ten years and that it was the best thing since sliced bread. They could get real relief, therefore it can't have done what you said. It was at that level. Epidemiologists, however, by and large were supportive.
>
> When my paper was published in the *American Review of Respiratory Disease*, the Medical Director of Riker wrote a letter in the *New England Journal of Medicine* criticising my work. It was a strange letter, not relating to anything that had appeared in that journal, but only relating to my work which had appeared in the *American Review of Respiratory Disease*. I had no chance to see it before it was published, and it attacked me as a publicity hound . . .
>
> I called the editor of the *New England Journal of Medicine*, and asked why the journal had published that letter when it concerned an article that wasn't published in the journal. He came back and said that he couldn't account for it . . . but that he felt badly about it and that he would let me publish a rebuttal . . . I said that I would rather that he did the following: ask two or three scientists, one an epidemiologist . . . to review this whole issue and write an editorial on the topic. He agreed to do this – but then he never did it . . .

After that, I wrote a letter to the Medical Director of Riker saying if he was going to continue to write letters like that I intended to do the following:

(1) ask the watchdog committee of the Food and Drug Administration to find out why a five times stronger preparation was licensed in these eight countries, but not in the United States;

(2) request all the information concerning the animal and human studies for safety and toxicity (because I knew they didn't have any);

(3) ask several reporters to do a story on it.

And they stopped, they got off my back right away.[32]

Letters criticising Stolley's work also appeared in the *American Review of Respiratory Disease*. He responded:

I have been struck with the reluctance of some pharmacologists to accept the conclusions of the brilliant epidemiologic investigations carried out by Inman and Adelstein, and Speizer and Doll (as well as others) attempting to explain this most dramatic and severe rise in asthma death rates. They usually return to the laboratory investigations to stress why these [inhalers] are harmless, but they have no plausible alternative theory to advance.[33]

MEDICAL SCEPTICISM

Stolley's work cleared up the major anomalies in the general *beta agonist hypothesis*, such as the absence of an epidemic in the United States. Some anomalies remained in the time-trend data for the countries in which isoprenaline forte was marketed, but these were only minor. For example, Gandevia (an Australian asthma specialist) had found no association between beta agonist inhaler sales and asthma deaths in Australia when sales were examined state by state.[34] However, Campbell (also an Australian asthma specialist) subsequently re-analysed the same data and found that there was a strong association in each of the four most populated states until 1966.[35] After that time, widespread publicity about the epidemics of deaths and the elimination of over-the-counter availability of beta agonist inhalers was followed by a fall in the death rate, even though total sales of isoprenaline forte did not change very much. A similar pattern had been seen in the United Kingdom.

These minor anomalies in the epidemiological evidence emphasise the limitations of time-trend data. Asthma deaths are caused by many different factors. Death rates can vary from year to year because of chance fluctuations (in a small country like New Zealand even a few extra deaths can make a difference in the national death rate), or changes in risk factors for asthma attacks (such as colds, house dust mites and tobacco smoking), asthma management, asthma education and safety warnings. Thus, although the overall pattern was clear, and the epidemics were closely linked to the introduction of isoprenaline forte, the death rates still jumped around from year to year. There was not a perfect fit between the sales of isoprenaline forte and the asthma death rates.

For these reasons, the scientists on both sides of the isoprenaline (forte) debate were agreed that the best way to resolve the issue would be to do a case-control study. For example, Stolley noted that 'Ideally, the most conclusive manner of investigating this question would entail a retrospective study of the products used by young asthmatic patients dying suddenly and unexpectedly at the peak of the epidemic'.[36] This would have involved studying the people who had died during the epidemic (the cases) and comparing them with a sample of people who had asthma during that time but who did not die (the controls). The medicines that they had been prescribed would then be ascertained for both groups. Although Doll had started to do a case-control study in the London area, it was not completed because warnings about the safety of isoprenaline had been issued and the epidemic had declined before there was time to finish the study.

Stolley tried to get the drug taken off the market in the United Kingdom. He later recounted:

However, a very famous pharmacologist said that he had seen some dog data . . . showing that dogs were inhaling very large amounts, and they weren't getting arrhythmias, and he just didn't think it was possible . . . My analysis was based on millions of people, looking at death rates, sales, and so on, and trying to put it all together in terms of what happened out there in the population . . . and I couldn't get [the drug] off the market because of six dogs . . . and because they didn't know how important [the epidemiological data] was.[37]

An additional problem was that clinicians and researchers trained in the methods of randomised trials are usually sceptical of epidemiological studies of exposures which are not randomised. Furthermore, doctors are trained to treat individual patients, and usually do it very well, but most do not understand the complexities of epidemiological studies of large numbers of people. Thus, the scepticism of respiratory physicians towards the epidemiological findings reflected their lack of understanding of the field as much as their psychological reluctance to accept the findings.

Nevertheless, Stolley's work 'rescued' the original work of Speizer, Doll and others. Although a case-control study of asthma deaths was never done, the findings for isoprenaline forte were so striking that isoprenaline forte became established as the major explanation for the 1960s epidemics of asthma deaths. Death from bronchodilator aerosols was eventually considered to be one of the most important adverse drug reactions since thalidomide,[38] and an editorial in the *British Medical Journal* entitled 'Asthma deaths: a question answered' concluded that Stolley's work 'strengthens still further the case for relating the increase in asthma deaths to aerosols and the sympathomimetic drugs they contain . . . the dose should not be increased in the absence of a normal response'.[39]

CHAPTER 2 History is rewritten

They rewrote history, they really did. It's like what happened in the Soviet Union – Trotsky was never born and played no part in the revolution. – PAUL STOLLEY[1]

The first time I heard the story of the 1960s epidemics was when Richard Doll was visiting New Zealand in 1981 and spoke at a special symposium in Dunedin. He was introduced by David Skegg, a former student of Doll's who had become the Professor of Preventive and Social Medicine at the University of Otago. Skegg highlighted some of Doll's remarkable achievements, which included such discoveries as the links between smoking and lung cancer, and between asbestos and lung cancer. In reply, Doll said that it was embarrassing to receive such praise, but that he understood that it was safe to receive such a tribute 'provided one does not inhale'. He then presented the story of the 1960s asthma mortality epidemics, noting that they were sudden and unexpected, that beta agonist inhalers had been established as the cause, and that the epidemics had receded when warnings had been issued and over-the-counter sales banned.

I heard a very different version of this story from respiratory physicians when I became involved in asthma research in 1988. In many different parts of the world, I was told that it had previously been thought that the 1960s epidemics were due to overuse of beta agonist inhalers 'but this has now been proved wrong'. When I asked how the beta agonist hypothesis had been disproved, or what alternative theories might explain the epidemics, I was met with blank stares. Everyone 'knew' that the hypothesis had been disproved, but no one knew how, and to question this 'established knowledge' would not make you popular. In fact, no new evidence had appeared that would justify claims that the role of isoprenaline forte in the epidemics had been refuted. The only relevant new data were published by Stolley in 1978,[2] and strengthened his original conclusions.

This process of 'reinterpretation' of the 1960s epidemic was due largely to the reluctance of doctors to consider that their treatment may have been dangerous. Their reluctance was encouraged by the minor anomalies in the time-trend data for the 1960s, which were emphasised and exaggerated in subsequent reviews. In addition, other factors, such as delays in seeking medical help, began to receive greater emphasis, and it was argued that 'errors of omission are a much more frequent cause of death than errors of commission'.[3] By 1979, a remarkably confused editorial in the *Lancet*, one of the world's top medical journals, could argue:

> There is a growing realisation that pressurized aerosols were
> probably not the main culprit, although excessive use of
> isoprenaline, as opposed to other constituents of pressurized
> aerosols, may have been an important contributory factor.
> With the decline in evidence that pressurized aerosols were
> responsible, attention has switched to the occasional failure of
> clinicians to treat asthma attacks promptly and adequately.[4]

This initially implies that isoprenaline forte was an important factor in the epidemic, but then discounts this possibility and instead emphasises the role of delays and 'undertreatment'. The *under-treatment hypothesis* suggested that the problem was not overuse of beta agonist inhalers; instead, it was underuse of other forms of treatment, such as inhaled corticosteroids, or emergency oxygen and oral corticosteroids for people with very severe asthma. The undertreatment hypothesis had some plausibility, since beta agonist inhalers relieve asthma symptoms and could therefore lead to delays in seeking help and being prescribed inhaled corticosteroids, which actually treat the underlying the disease (by reducing inflammation in the airways) rather than just treating the symptoms (bronchospasm). However, the undertreatment hypothesis was essentially just another version of the delay hypothesis (see

Chapter 1), as delays in starting appropriate treatment could be caused by overuse of beta agonist inhalers. Furthermore, if patients continued to delay seeking help, they would continue to take their beta agonist while becoming more hypoxic, and acute toxicity was therefore more likely to occur.

It is also very difficult to see how treatment delays could have spontaneously caused the 1960s epidemics of deaths in some countries, which just happened to be those in which isoprenaline forte was marketed, but not in others. Why would patients suddenly start delaying seeking help? This would not happen spontaneously, but could be caused by overuse of a powerful drug like isoprenaline forte that relieved their symptoms. Moreover, if undertreatment was the major problem, then it is also difficult to explain why subsequent increases in the range of asthma drugs available and the quantity of drugs prescribed in the 1970s and 1980s did not lead to a fall in asthma deaths. Thus, undertreatment is not a plausible explanation for the 1960s epidemics, except in the sense that overtreatment with high-dose beta agonists may have led to delays in seeking more appropriate treatment in hospital.

Despite the limitations of the undertreatment hypothesis, it became the most commonly accepted interpretation of the 1960s epidemics, and the role of beta agonists was discounted. By 1988, a 'state of the art' review was arguing that 'There is no such thing as "excessive" use of beta-2 agonists . . . these drugs should never be withheld for fear of toxicity'. [5]

ORCIPRENALINE AND THE FDA

The tendency to discount the role of isoprenaline forte in the 1960s epidemics gathered strength as a result of the controversy that followed the decision of the United States Food and Drug Administration (FDA) to license orciprenaline (another beta agonist) for non-prescription sale. Although isoprenaline accounted for most of the sales during the 1960s epidemics, orciprenaline ('metaproterenol' in the US) accounted for 8 per cent of aerosol sales during the epidemic in England, and a study of asthma deaths in England during the epidemic found that about 25 per cent had been prescribed orciprenaline.[6] The drug was also associated with reports of asthma deaths in Australia, where it had nearly 50 per cent of the market in some states in the mid-1960s.[7] Orciprenaline has similar side-effects on the heart to isoprenaline, but it was sold in a regular dose and was not marketed in a high-dose preparation like isoprenaline forte.[8]

An FDA decision to make orciprenaline available for non-prescription sale in the United States in 1983 was overturned two months later, because of renewed concerns about the role of such drugs in the 1960s epidemics. A subsequent editorial in the *New England Journal of Medicine* emphasised that

> Inhaled [orciprenaline] is safe and generally effective at recommended doses ... For patients who have chosen to self-medicate, however, the rapid and readily perceived degree of improved comfort [may cause] the patient to delay seeking medical care ... Under these circumstances, death may occur because overdose causes ventricular arrhythmia, but it is more commonly speculated that airway obstruction unresponsive to beta agonists progresses because of drug tolerance or mucosal secretions and edema from the asthma.[9]

The conclusions of this editorial were challenged by William Wardell, the United States Medical Director of Boehringer Ingelheim, the company which manufactures orciprenaline (the same company manufactures fenoterol, which will appear later in this story). He argued that over-the-counter availability had been ruled out as a factor in the increased death rate in the 1960s. In justification of this claim, he referenced the paper by Inman and Adelstein[10] which had actually reached the opposite conclusion. Wardell also cited the Australian studies of Gandevia (discussed in Chapter 1) and concluded that 'In the face of these facts, it is scientifically impossible to cling to the theory that the transient rise in asthma-related deaths in the United Kingdom and Australia was due to excessive and unsupervised use of metaproterenol aerosols resulting from nonprescriptions sales.'[11]

In response, Weinberger and Hendeles pointed out that the anomalies in the time-trend data cited by Wardell were relatively minor, and were probably due to widespread publicity about the epidemic in Australia in 1966. They concluded that 'It is not only possible but prudent to "cling to the theory" that deaths have previously resulted from unsupervised used of aerosol bronchodilators and to consider the possibility that history could be repeated by nonprescription availability of these medications, accompanied by intensive indirect marketing.' [12]

THREE REVIEWS

Following this controversy, two reviews were published which questioned the role of beta agonists in the 1960s epidemics. One review was written by Stephan Lanes and Alex Walker, from Epidemiology Resources Inc, an epidemiological consulting firm that regularly does work for industry.[13] This review was commissioned by Boehringer Ingelheim, the manufacturer of orciprenaline, and was published in the *American Journal of Epidemiology* in 1987[14] with an acknowledgement of their funding from Boehringer Ingelheim. The review emphasised the minor anomalies in the general beta agonist hypothesis. The review also discussed Stolley's more specific version of the beta agonist hypothesis that focused on the role of isoprenaline forte, but dismissed the striking time trends in the six countries where isoprenaline forte was sold heavily, and instead emphasised the minor anomalies in the time-trend data. The review presented no alternative hypothesis to explain the epidemics of deaths, other than to propose that 'asthmatics die from asthma' (which no one was disputing) and that the epidemics may have been due to changes in the incidence of acute asthma. The latter hypothesis had already been considered and rejected by Speizer, Doll and others who had concluded that it was very unlikely that the incidence of acute asthma, after being stable for nearly a century, would suddenly double or triple within a few years and then decline again in several selected countries, but not in others, and that this 'spontaneous' epidemic would coincidentally parallel the sales of isoprenaline forte.[15]

A similar review, which cited the FDA decision on orciprenaline as its justification, was written by Alvan Feinstein's group at

Yale University. Feinstein is somewhat of a legend among epidemiologists (and lawyers) because of his strident attacks on epidemiological research.[16] This review was published in 1987 in *Archives of Internal Medicine*,[17] but the source of funding was not mentioned. This paper not only documented some of the anomalies in the beta agonist hypothesis, but also raised an alternative hypothesis that the epidemics may have been due to diagnostic exchange, in that deaths from respiratory infections may have been misclassified as deaths from asthma. This alternative hypothesis had also been considered and rejected previously by Speizer, Doll and others. The paper did discuss some of the shortcomings of the *diagnostic exchange hypothesis*, and noted that it was very difficult to see how such a bias could have suddenly occurred and then reversed in several selected countries but not in others. Nevertheless, the general message of the paper was that the link between the epidemics and beta agonists had been refuted and that even the occurrence of the epidemics was not proven.

Paul Stolley was also asked by Boehringer Ingelheim to write a review of the 1960s epidemics. As he later recounted:

> I was hired by Boehringer Ingelheim, and actually accepted a fee. They wanted me to . . . say that Alupent (orciprenaline), which is the drug they were trying to get sold over the counter in the United States, was not the cause of the epidemic, that the cause was [isoprenaline].[18]

There is a large market for medicines which can be purchased at a pharmacy without a doctor's prescription. At the time, adrenalin was the only asthma drug available without prescription, but Boehringer had almost persuaded the FDA to approve orciprenaline for non-prescription sale. Stolley thought that orciprenaline was relatively safe because the epidemics had been caused by isoprenaline forte and there was no evidence that orciprenaline was a problem.

> They paid me a very nice fee. I went to a meeting in Boston and made a presentation to the Academy of Allergists. I

was then amazed to find that Boehringer Ingelheim were commissioning a group to say the epidemic hadn't happened, then another group to say that maybe the epidemic happened but that you can't use these kind of statistics to work anything out.

So they had three lines of defence. One was that the epidemic hadn't happened; the second was that it happened but we don't know what caused it; and the third was me, who said the epidemic happened, and we know what caused it, but it wasn't Alupent . . .[19]

As Stolley saw it:

What you have is a company manufacturing these drugs, believing that they're useful, that they work. They want the over-the-counter market, they think they have a safe drug, they meet resistance, and they try to get around it. The one thing they never think of is 'can this really be true?', and when they read things that are adverse, they hire people with predictable views, by and large.

So, when I found that out, of course, I stopped my relationship with them.[20]

Despite the obvious shortcomings of the reviews commissioned by Boehringer Ingelheim, and their lack of plausible explanations for the epidemics, they fostered a belief among doctors that the epidemics were not caused by beta agonists. For example, one review published in the *Medical Journal of Australia* in 1990 cited the review by Feinstein's group as justification for claiming that 'The suggestion that the so-called epidemic of asthma deaths in the 1960s was due to increased usage of isoprenaline inhalers has largely been refuted, and changes in diagnostic labelling of asthma now seems to be a likely explanation'.[21]

Similarly, a paper by Benatar (a respiratory physician in South Africa), published in the *New England Journal of Medicine* in February 1986, claimed that 'For many years the hypothesis that sudden death from asthma resulted from the use of aerosol bronchodilators was widely believed within the medical profession and the lay public, with a consequent reluctance to prescribe or use aerosol bronchodilators . . . subsequent studies refuted these reports.'[22] However, the references that Benatar cited all related to the safety of the fluorinated hydrocarbon propellants rather than the safety of the beta agonists themselves.* The only other reference in support of Benatar's claim was to a review which made the same claim, but without producing any evidence.

Stolley submitted a letter on this point to the *New England Journal of Medicine*, which rejected it, but he subsequently managed to have the letter accepted by *Annals of Internal Medicine*. In the letter, he drew an analogy with the story of the Bellman in *The Hunting of the Snark* by Lewis Carroll, that 'what I tell you three times is true'. In other words, if a falsehood is repeated often enough, then people begin to believe it.[23]

Stolley later observed that the suggestions that the epidemics were due to poor medical management, rather than the medicines themselves, were inconsistent with the fact that the epidemics happened all over the world at the same time, in some countries and not in others:

It is so implausible to suggest that doctors suddenly became careless in certain countries and not others, beginning within a few months of one another. That just makes no sense at all. In Norway they had the epidemic, in Sweden and Denmark they didn't, the United States had no epidemic, Canada had no epidemic. It began almost simultaneously in New Zealand, Australia, England and Wales, Scotland and Ireland. To suggest that this is all due to poor medical management, you would have to postulate some neurosurgeon going round doing lobotomies on doctors simultaneously. It just makes no sense. Common sense tells you that, and yet this [became] the common explanation.[24]

---

* At one time it was considered that fluorinated hydrocarbon propellants could be the cause of the epidemics, but this explanation was ruled out at an early stage because the propellants were used in virtually all countries, not just in those which had epidemics.

Those who cannot remember the past are condemned to repeat it
– GEORGE SANTAYANA[1]

In 1976 a second epidemic of asthma deaths began. This time it occurred only in New Zealand.

It was not identified for several years, because of the usual delays in the publication of the official death statistics. However, there was some concern about the high asthma death rate in New Zealand as early as 1978, when the Medical Research Council of New Zealand established the Asthma Task Force.

THE ASTHMA TASK FORCE

The task force was chaired by Tom O'Donnell, who had been the Professor of Medicine at Otago University's Wellington School of Medicine since its foundation in 1975. He became the dean of the school in 1986 and retired in 1992. The nucleus of the task force was O'Donnell and three other respiratory physicians: Peter Holst, a senior lecturer in O'Donnell's department; Harry Rea, a respiratory physician at Green Lane Hospital in Auckland; and Malcolm Sears, a senior lecturer in the Department of Medicine at Otago University in Dunedin.

At various times the task force included several other respiratory physicians, and also an epidemiologist, Robert Beaglehole, at that time a senior lecturer in the Department of Community Health at Auckland Medical School. He was a specialist in cardiovascular epidemiology, and was internationally recognised as an epidemiologist. Although he had little experience of asthma epidemiology, neither did anyone else in New Zealand, and so he agreed to become involved in the task force's work.

The first public report of the epidemic of deaths was published in 1981, but not by the Asthma Task Force. Instead, it came from Doug Wilson, Professor of Immunology at Auckland Medical School. He had been contacted by a member of the Auckland Asthma Society Committee, who was worried that three members of the committee had died of asthma during the previous year. Wilson reviewed the records of a series of 22 asthma deaths in the Auckland area, and published the findings in the *Lancet* on 6 June 1981. Reporting an apparent increase in young people dying suddenly from asthma in Auckland, he noted that

> In 16 patients death was seen to be sudden and unexpected. Although all were experiencing respiratory distress, most were not cyanosed and the precipitate nature of their death suggested a cardiac event, such as arrest, inappropriate to the severity of their respiratory problem . . . We feel that the only reasonable explanation is that the change must be a reflection of changes in the patterns of treatment of asthma in Auckland.[2]

In subsequent media reports, Wilson emphasised that the patients were seen to be using 'plenty of drugs' and that underuse of asthma medicines did not seem to be the explanation for the deaths.[3]

The paper suggested two hypotheses to explain the new epidemic. One was essentially the drug-induced delay hypothesis, that 'when regular symptomatic treatment with bronchodilators replaces inhaled steroids or cromoglycate the patient's perception of an asthma crisis may be delayed'.

A second hypothesis, which received more attention, was an interesting variant of the toxicity hypothesis. Wilson noted that new oral theophylline products had recently become available, and that the idea of taking oral theophylline together with beta agonists had been promoted in New Zealand. Oral theophylline is not a beta agonist but another type of bronchodilator which is taken in tablet form rather than by inhaler. Wilson hypothesised that 'there may be an additive toxicity between theophyllines and inhaled beta-2 agonists at high doses which produces cardiac arrest'.

This hypothesis was interesting because it seemed to account for an apparent paradox. It seemed reasonable to suggest that the second New Zealand epidemic could have been caused by beta agonists, since isoprenaline forte had been the cause of the first epidemic. However, sales of isoprenaline had fallen off dramatically, and a new generation of beta agonists had been marketed. These new drugs were believed to have far fewer side-effects on the heart than isoprenaline. Furthermore, these drugs were thought to have been widely used in other countries, but only New Zealand had experienced an epidemic of asthma deaths. Wilson's hypothesis appeared to resolve this contradiction, because he suggested that what was unique about New Zealand was that beta agonists were being used together with oral theophyllines, and a drug interaction could be causing the deaths. It turned out that there was a weak interaction of the type he described,[4] but that this was not strong enough to account for the epidemic.

Wilson's hypothesis was phrased entirely in terms of a 'class effect' of beta agonists. He assumed that all of the beta agonists were the same and that they would carry similar risks. He did not consider the possibility that there could be differences between the three beta agonists in common use: salbutamol (Ventolin), fenoterol (Berotec) and terbutaline (Bricanyl).

Wilson's paper received huge media coverage, and generated massive publicity about the existence of a new asthma mortality epidemic in New Zealand. In a television interview, Wilson argued:

> We are raising this [hypothesis] as a possibility now, because with the rate at which patients are dying, about one every three or four days, it is obviously urgent for us to raise this as an issue so that some caution may come into our approach to the situation. If we're wrong, it doesn't matter. If we're right, then obviously we've saved some lives.[5]

However, there was great hostility to Wilson's views from the respiratory physicians. Not only had his report pre-empted the work of the task force, but had once again raised the possibility that asthma drugs were causing deaths, when it was widely believed, and hoped, by doctors that this old idea from the 1960s had been laid to rest. Not only was Wilson's hypothesis an affront to New Zealand respiratory physicians, but there was concern that the publicity about his paper might cause panic – there were reports of patients throwing away their asthma medicines. However, the asthma death rate actually fell following the media publicity.[6] Similar falls in the death rate had occurred in the 1960s after the epidemics were identified and the possibility that asthma drugs might be responsible was publicised.[7]

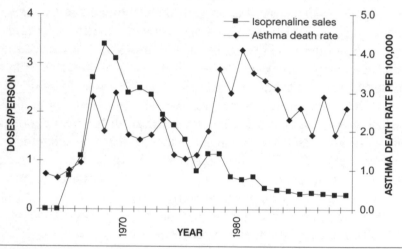

FIGURE 4: New Zealand isoprenaline forte sales per capita and asthma deaths per 100,000 in the 5- to 34-year age-group between 1962 and 1988

ROD JACKSON

One of the few researchers who seriously considered that the epidemic might have been caused by an asthma drug was Rod Jackson, at that time a PhD student in epidemiology with Robert Beaglehole in Auckland. He analysed national data on asthma deaths, and found that the asthma death rate (in the 5- to 34-year age-group) had been about 1.4 per 100,000 in 1975, but had increased to about 2.0 in 1976, and had leapt to 3.6 in 1977, before peaking at 4.1 in 1979 (see Figure 4). Thus, the asthma death rate had doubled or tripled in the space of a few years. This was certainly a new epidemic, which was even more sudden and dramatic than the first one. It was not explained by isoprenaline forte, because the sales of that drug had fallen almost to nothing (see Figure 4).

Jackson met Richard Doll on his New Zealand visit in 1981, and asked if he could discuss his data in the light of Doll's work on the 1960s epidemics. Jackson later commented: 'He took one look at the graph, and placed his finger on it, saying "that's iatrogenic".'[8] In other words, Doll was saying that the epidemic must have been caused by medical treatment.

Jackson, together with several colleagues, including Robert Beaglehole and Harry Rea (from the Asthma Task Force), prepared a report on the epidemic. They considered the possible explanations for the epidemic, using methods similar to those used to study the 1960s epidemic in England and Wales. As with the 1960s epidemic, they concluded that the new epidemic in New Zealand appeared to be real, and could not be explained by changes in the classification of asthma deaths, inaccuracies in death certification or changes in diagnostic fashions. They concluded that it was very unlikely that the epidemic could be due to changes in the prevalence of asthma in New Zealand, and that the most likely explanation, as with the 1960s epidemics, appeared to be an increased case-fatality rate related to changes in the management of asthma in New Zealand.[9]

A subsequent editorial in the *New Zealand Medical Journal* concluded that 'Undoubtedly New Zealand has a real epidemic of asthma deaths for which an urgent explanation is needed. There is nothing to suggest that asthma in New Zealand is any different to that elsewhere except for its propensity to cause death.'[10]

Once again, the most likely explanation was that the epidemic was due to the introduction of a new drug. However, no one knew what the drug could be, since isoprenaline was no longer widely used, and the new beta agonists seemed to be safe.

IAN GRANT'S VISIT
The controversy was further inflamed by the visit to New Zealand in September 1982 of Professor Ian Grant from the University

of Edinburgh. Grant travelled around New Zealand as a guest of the New Zealand Asthma Foundation, and spoke at a series of well-attended public meetings organised by asthma societies. On his return to Scotland, Grant published his thoughts on asthma in New Zealand in the *British Medical Journal*.[11] He noted that in many respects the use of asthma drugs seemed similar in New Zealand and the United Kingdom. However, New Zealand differed from the United Kingdom in that home nebulisers were widely available. These are machines with an electrically operated pump and a face-mask, and can administer much larger doses of beta agonists than can be obtained from an inhaler. He suggested that the widespread use of high doses of beta agonists delivered by air-driven home nebulisers in New Zealand could be a factor in the epidemic, particularly by causing patients to delay seeking medical help. Grant argued:

> Respiratory physicians in the United Kingdom have so far adopted a much more cautious attitude towards providing electrically operated air compressor nebulisers for [use by] patients in their homes. They fear that this form of treatment may reproduce all the hazards associated with the unbridled use of isoprenaline inhalers in the 1960s . . .
>
> The situation in New Zealand is radically different. In the past two years 6000 nebulisers have been purchased by asthmatic patients without necessarily having the approval of their doctors . . . when airflow obstruction is so severe that massive doses from a nebuliser are required the patient ought to be in hospital. I had the impression, which was confirmed by most of the respiratory physicians I met in New Zealand, that the popularity of nebulisers was a fashion encouraged by pharmaceutical companies and the manufacturers of nebulisers which the medical profession had so far failed to counter . . .
>
> It is not inconceivable that we may already be witnessing the beginning of a 're-run' of the international epidemic of deaths from asthma in the 1960's . . . The parallels are too dangerous to ignore . . . Patients are . . . being given . . . uncontrolled access to a form of treatment from which they often derive undoubted

subjective benefit but which on occasion may conceal the danger of the disease itself and cause them to underestimate the importance of seeking expert medical advice at times when their lives are potentially at risk.

Grant's report contained several errors, particularly in the availability of nebuliser solution; it was available only on prescription, but he thought that it was available over the counter. More importantly, Grant apparently did not realise that the epidemic of deaths had started in 1976, whereas the increase in nebuliser sales began only in 1980. Thus, it seemed very unlikely that the home nebulisers could account for the epidemic.

Nevertheless, Grant correctly identified the complacent attitude to beta agonists (both in inhalers and in home nebulisers) in New Zealand. Furthermore, he pointed out the importance of hypoxia as a cause of asthma death, noting that high doses of beta agonists could make hypoxia worse, and that therefore it was potentially dangerous for patients with a severe asthma attack to use home nebulisers without a supply of oxygen. Grant also pointed out that the New Zealand system of primary health care could exacerbate the overuse of beta agonists in New Zealand: 'I wondered . . . if the method of payment by partially recoverable fees for each item of service might inhibit asthmatic patients from consulting their doctors as often as they should and perhaps create a disincentive to earlier referral to a hospital physician.' Thus, overuse of beta agonists was very likely to occur when there was general complacency about the dangers of these drugs, and when there were financial barriers to visiting a general practitioner, but beta agonists were available free through hospital doctors or repeat prescriptions.

Grant further explained his concerns in a subsequent letter to the Asthma Task Force:

I think that you and your colleagues still tend to underestimate the terrifying (to me) false sense of security conferred by the mere possession of nebulisers . . . The risk, as I see it, is that every patient who had a nebuliser . . . knew from experience that it usually controlled an asthmatic attack, and was certain to have used it at least a few times before calling a doctor. In other circumstances there would have been a good response, perhaps 99 times out of 100 (which is of course why nebulisers are so popular with asthmatics), but in a severe attack there must be an almost irresistible temptation to use the nebuliser 'just once more'. That, I am convinced, is how fatal delays in starting more effective treatments can occur.[12]

Grant's report was unpopular, to say the least. There was widespread indignation among respiratory physicians that this Scotsman had toured the country, accepted their hospitality, and then written an article in one of the world's most widely read medical journals saying that the epidemic was occurring because New Zealand doctors were giving substandard treatment and being irresponsible in handing out dangerous drugs. Two members of the Asthma Task Force expressed their indignation in a letter to the *British Medical Journal*:

We were disappointed to read Dr I W B Grant's account of asthma in New Zealand . . . The paper contains errors of fact and emphasis and tends to leave the impression that Dr Grant has been able to identify a number of problems which had escaped the notice of local doctors dealing with bronchial asthma. This is regrettable and misleading. The few facts that are available were made known to Dr Grant during discussions with doctors in New Zealand.[13]

Grant replied:

I was fully aware that my article would provoke controversy, but I did not expect the reaction it elicited from Dr D.C. Sutherland and Dr H.H. Rea . . . [who] accuse me of propagating speculative ideas poached from my medical hosts. If informed speculation is so much to be deplored I would now expect them to condemn the British doctors who warned the medical profession about the dangers of isoprenaline inhalers in the late 1960s . . . I feel sad that what was meant to be a genuine, although perhaps incautious, attempt to present an unbiased view on what is obviously a major health problem in New Zealand should have caused so much offence. I am, however, completely unrepentant.[14]

This was the start of a lengthy series of letters in the *British Medical Journal* and the *New Zealand Medical Journal*, debating Grant's claims and expressing irritation at his actions. For example, two Christchurch specialists wrote:

Dr Grant detected anxiety in his New Zealand audiences after public meetings. This is quite understandable. In the last year an American respiratory specialist while visiting Christchurch blamed death in asthma on the [beta agonist inhaler], which he likened to a loaded revolver. A report from Auckland which received wide coverage in the news media suggested an association between asthma deaths and the combined use of theophylline and [beta agonist inhalers]. Now Dr Grant, receiving equally wide coverage in the news media, suggests that nebulisers are dangerous. The urge to blame medication for asthma deaths may be misplaced . . . These frequent alarming opinions based on the flimsiest of evidence can do nothing but undermine the patient's confidence in his drugs and erode the morale of his doctor.[15]

Grant's paper was reprinted in the *New Zealand Medical Journal*, with an editorial by Tom O'Donnell, Chair of the Asthma Task Force, who argued:

One is inclined to be sensitive to the views of a visitor who after a brief stay expresses his or her opinions which are critical of local practices in a field within one's own area . . . Dr Grant accepted the invitation to come and talk here. During his visit he recognised the problem of asthma, and has sought to identify aspects of management differing from his own practice. We must be grateful to him. Our own national asthma mortality survey should provide information regarding his suspicions. How correct is his suspicion regarding the unsupervised use of bronchodilator drugs through powered nebuliser units, the lack of stressing a need for oxygen therapy in severe asthma, or the risks with the greatly increased use of sustained release theophylline preparations? We will be tempted to dismiss these views . . . The visit of Dr Grant will have been successful if we realise the potential seriousness of asthma, attempt to improve the understanding of asthma by patients, try to reach the many asthmatics in our population and facilitate prompt access to the most effective therapy and supervision according to individual needs.[16]

Thus, the Asthma Task Force survey of deaths was already being promoted as the answer to the criticisms of the overseas experts. Furthermore, before the survey had even been completed, problems of patient education and access to therapy, rather than the hazards of beta agonists, were already being promoted as likely explanations for the epidemic of deaths.

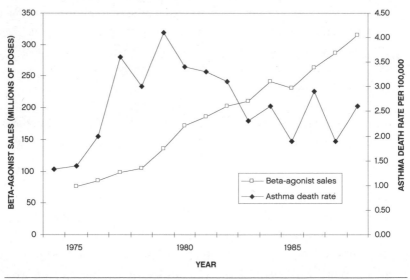

FIGURE 5: New Zealand beta agonist sales (millions of doses) and asthma mortality per 100,000 in the 5- to 34-year age-group

The hypotheses of Wilson and Grant both involved 'class effects' of beta agonists. They assumed that all of the beta agonists had the same side-effects, and that if there was a problem with safety then it would apply to all of them. These hypotheses were called into question, even bearing in mind the limitations of time-trend data, by subsequent research by Gay Keating and Rod Jackson in Auckland.[17] This showed that the epidemic had started in 1976, but that sales of beta agonist inhalers had begun to increase markedly only in 1979 (see Figure 5), and that sales of nebulisers had begun to increase only in 1980. They concluded that 'Although the temporal association between mortality and sales of drugs suggests that direct drug toxicity is unlikely, there may be more subtle adverse effects of drug use'.

Nevertheless, the hypotheses of Wilson and Grant resulted in a great deal of publicity and concern about the safety of asthma drugs in New Zealand and about the dangers of overuse; the death rate began to fall (see Figure 5), just as it had when similar warnings were issued during the 1960s epidemics.

THE ASTHMA TASK FORCE SURVEY

While all this was happening, the Asthma Task Force was continuing its work. It had started a study of all asthma deaths in people under 70 years of age in New Zealand during the 2-year period August 1981 to July 1983. However, the study did not include a control group, and (with one exception, discussed below) the findings for individual asthma drugs were not reported. This was surprising, given that the main concern was the possible role of asthma drugs, and that it had previously been agreed that a case-control study would have been the most appropriate way to investigate the 1960s epidemics.[18]

It is not clear why the survey never included a control group. This may have reflected a lack of epidemiological expertise in the core members of the Asthma Task Force. It probably also reflected a failure even to recognise the possibility that a particular drug within the range of asthma drugs being used could have serious adverse effects.

Like Robert Beaglehole and Rod Jackson, Paul Stolley had urged that the New Zealand study should include a control group:

> I had correspondence with people in New Zealand as far back as 1978. They said they were having another increase in asthma [deaths] and they asked my advice because of my previous work, and they showed me some graphs. And I said, 'You know, changes like this are usually due to a common exposure and that's likely to be a drug, and a case-control study should answer the question'.
>
> There was lots of money available, but they didn't spend it that way. Instead they did a case series without controls. It's strange, it really is strange. And yet these are the same people who won't believe what happened in the 1960s because there were no case-control studies.[19]

At the urging of Robert Beaglehole and Rod Jackson, a control group was included in a sub-study of the deaths in the Auckland area.[20] However, the Auckland case-control study involved only 44 deaths, and the findings for individual asthma drugs were not reported when the paper was published in 1986, because it was considered that the numbers were too small to analyse properly.

The first report of the 271 asthma deaths analysed in the national study was published in 1985. The report did not specify any criteria for overuse, but concluded that 'excessive use of bronchodilator drugs did not account for the high mortality rate . . . [because only] nine [of 271] cases were identified where there could have been excess usage of beta-2 agonist before death'.[21]

The report instead emphasised problems of underuse of preventive treatment, particularly corticosteroids.* However, there

---

* It is true that corticosteroids are the best preventive treatment for chronically severe asthma. Oral corticosteroids (taken as tablets) are used for very severe asthma, whereas inhaled corticosteroids (taken with an inhaler) are used for moderately severe asthma. These preventers work slowly and need to be taken regularly. This is a major problem, because they do not provide the instant relief obtained with 'relievers' (including the beta agonists). As a result, many people do not take their preventer medicines properly, or do not take them at all.

was little change in prescribing patterns for inhaled corticosteroids at the time the epidemic began (as shown in Figure 6). Further-more, the epidemic occurred only in New Zealand, and not in other countries, most of which had lower prescribing of inhaled corticosteroids than in New Zealand. Thus, underuse or underpre-scribing of corticosteroids could not explain why undertreatment should have suddenly caused a tripling in the death rate between 1975 and 1977 in New Zealand, but not in other countries.

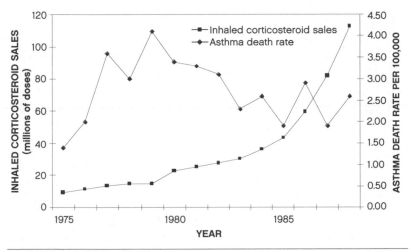

FIGURE 6: New Zealand inhaled corticosteroid sales and asthma mortality per 100,000 in the 5- to 34-year age-group

It appears that there were considerable disagreements within the Asthma Task Force, particularly between Beaglehole and the respiratory physicians. In contrast to the initial dismissal of any possible role of overuse, a subsequent report noted that 'Most patients whose final episodes had lasted for several hours had re-peatedly used their inhaled bronchodilator aerosol, and in some cases a domiciliary nebuliser, to administer large doses of beta-agonist, without seeking additional therapy with corticosteroids'.[22] A subsequent paper written by Beaglehole argued that 'The most likely explanation for the epidemic in New Zealand is that an ex-cessive reliance on modern bronchodilator therapy resulted in delays in implementing appropriate treatment for acute progressive attacks . . . [but] . . . the data does not exclude the possibility of a direct toxic effect in a susceptible subgroup of patients'.[23]

At least some other New Zealand doctors were also sceptical about the Asthma Task Force's report. Mackay and van der Westhuyzen, both from Nelson Hospital, asked in the *New Zealand Medical Journal*: 'Why is it that despite the massive increase in knowledge, education and drug expenditure on asthma in the last 20 years, the morbidity and mortality from this complaint has got worse? It certainly has not got worse through the inadequate deployment of corticosteroids!'[24]

Nevertheless, the initial dismissal of a possible role of beta agonists, along with the emphasis on problems of undertreatment with corticosteroids, was widely quoted. The 'fact' that overuse could have been a factor in only 3 per cent of the deaths was reported around the world as compelling evidence that the New Zealand epidemic had not been caused by overuse of beta agonists. The message in the media was that the main problem was underuse, rather than overuse, of asthma medicines.

Following this publicity in 1985, the asthma death rate, which had fallen to about 2.0 per 100,000 in the 5- to 34-year age-group following the publication of Wilson's paper in 1981, increased again to a level of about 3.0 per 100,000 in 1986 (see Figure 6). By 1988, Malcolm Sears was writing in the *New Zealand Medical Journal* that 'The problem is not resolved; it will not go away; we must continue to ask, and try to answer by carefully designed and controlled studies, questions about the nature of asthma in New Zealand and its optimal management'.[25] The hypothesis that the epidemic could have been caused by asthma drugs had been stifled, but the epidemic had not gone away.

CHAPTER 4 'Maybe these drugs are not all the same'

Dr Julian Crane started shaking and his heart raced. It wasn't a case of love at first sight. Dr Crane had taken a few puffs of the asthma inhalant fenoterol, and for the first time he really understood what his patients had been telling him.[1]

By the time the first Asthma Task Force publication appeared in 1985,[2] it seemed that the mystery of the second New Zealand epidemic would never be solved. The death rate had fallen a little since Wilson's paper had appeared in 1981 – from an average of about 3.5 per 100,000 at the peak of the epidemic in 1977–80 to an average of about 2.5 per 100.000 in 1983–85. However, despite this fall in the death rate, which was probably due to the media publicity about the dangers of overuse of asthma drugs, it was still by far the highest in the world. The next closest was in Australia, with a death rate of about 1.5 per 100,000.[3]

The Asthma Task Force was arguing that undertreatment was the main cause of the deaths, but there was no evidence that undertreatment or underuse of asthma drugs had suddenly become a serious problem in New Zealand when the epidemic started in 1976. In fact, New Zealand had the highest per capita use of asthma drugs in the world.[4] No other factor which could plausibly have caused the epidemic had been suggested. It still seemed most likely that the epidemic was due to some change in the management of asthma in New Zealand, and the most plausible explanation was that it was in some way related to asthma drugs, just as the 1960s epidemics had been. However, the hypothesis that it was due to beta agonist inhalers in general just did not seem to fit. The epidemic had started in 1976, but there had been very little change in total sales of beta agonists at that time, and sales had begun to increase only in 1979 when the epidemic had already peaked.

It thus seemed unlikely that the cause of the epidemic would ever be discovered, and only a few researchers felt that asthma drugs could be responsible. Rod Jackson and Robert Beaglehole were continuing to monitor the situation, and to explore the possibility that it was due to beta agonists in general, or to a dangerous combination of drugs such as beta agonists and oral theophylline. In the meantime, Doug Wilson, who had suggested the latter hypothesis, had moved on from his position as Professor of Immunology at Auckland Medical School and, after a spell in Saudi Arabia, had become Medical Director of Boehringer Ingelheim in New Zealand.

JULIAN CRANE

The mystery would probably never have been solved if it had not been for Julian Crane. He was employed as the Boehringer Ingelheim Research Fellow at the Wellington School of Medicine, and was working with Tom O'Donnell, the Head of the Department of Medicine and Chair of the Asthma Task Force. Crane was doing several clinical studies involving some of Boehringer's asthma drugs. In 1985 he was funded by Boehringer to present some of his research at a conference in Florence. On the long plane flight home he caught a cold, and when he got home it developed into wheezing. He had asthma symptoms for the first time in his life. He needed to take a beta agonist, and he had plenty of free samples of fenoterol lying around his office. This was the beta agonist that Boehringer Ingelheim had marketed in New Zealand under the brand-name of Berotec. Crane took a few puffs, and was struck by the effects. He started shaking and his heart raced. His patients had been telling him about these side-effects for years, but he had not really taken them seriously until they happened to him. He later commented: 'I became interested in the side effects and started comparing them with other asthma drugs. I was amazed at the differences between them'.[5]

46

He found that fenoterol seemed to have much greater side-effects than salbutamol (marketed under the brand-name of Ventolin) and terbutaline (marketed under the brand-name of Bricanyl), the other two beta agonists commonly used in New Zealand. He then asked a question which it seems no one had asked before: 'Maybe these drugs are not all the same?'

It was then that Crane remembered Doug Wilson's first report on the second New Zealand epidemic, which had appeared in the *Lancet* four years before.[6] While Wilson's paper had been about a possible class effect of beta agonists, and no individual beta agonists had been mentioned in the main text, in a table he had listed the individual drugs patients had been using. Of the 16 sudden asthma deaths, drug information had been reported for 13, and 10 of these had been using fenoterol. This seemed a very high proportion, since fenoterol had accounted for less than 30 per cent of the inhaled beta agonist market in New Zealand at the time. Crane had noticed this when the paper was first published in 1981 and had intended to write a letter to the editor, but he had never finished it. If anyone else noticed the high proportion of fenoterol users in the table, no comment on it has ever appeared in print.

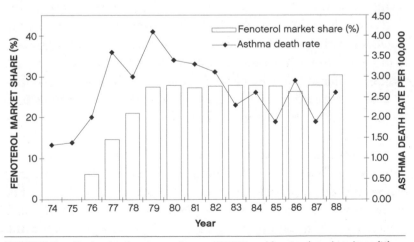

FIGURE 7: New Zealand asthma mortality per 100,000 and fenoterol market share (%)

Crane went downstairs to the medical school library and studied old issues of the *New Zealand Medical Journal*. Fenoterol had been introduced into New Zealand in 1976, and the epidemic had started in the same year. The drug had captured nearly 30 per cent of the market by 1979, and had held a constant market share ever since. The sudden increase in deaths in 1976–79 closely paralleled the sudden rise in its market share. Thus, while there was no correlation between total beta agonist sales and the rise in the death rate (see Figure 5), there was a strong correlation between the fenoterol market share and the rise in the death rate (see Figure 7).

Crane was even more surprised to find that the drug had never been licensed in the United States. The reason for this is not clear. Some claimed that it was due to the drug's known side-effects on the heart. Others claimed that it was because the drug had been found to cause cancer in mice,[7] and that this problem also occurred with some other beta agonists which had been marketed before fenoterol and before the problem was known. Whatever the explanation, the drug was not licensed in the United States and was not widely used in most other countries. In all other English-speaking countries, the market share was less than 10 per cent and no country had had such a dramatic increase in sales as New Zealand. Although the market share was higher in West Germany and several other European countries, the per capita use was only a third of that in New Zealand. Furthermore, although there had not been an epidemic of deaths in West Germany, the asthma death rate there nearly doubled in the five years after fenoterol was introduced.[8]

It is not clear why the marketing of fenoterol was so much more effective in New Zealand than in other countries, particularly when anecdotal reports of side-effects were so common. The promotion of fenoterol may have been easier because terbutaline, which was the second best-selling beta agonist in most countries (salbutamol was the first), had not been heavily promoted in New Zealand and there was room for fenoterol to capture the No. 2 market position.

The marketing of fenoterol may also have been helped because Boehringer Ingelheim was the major funder of asthma research in New Zealand, and enjoyed good relationships with the leading respiratory physicians, as indicated by the research fellowship which funded Julian Crane. Another indication of the links between the drug company and the asthma specialists was the series of annual asthma symposia in Rotorua. Funded by Boehringer Ingelheim, these brought together several invited overseas speakers and 30 to 50 of the leading asthma specialists in New Zealand for a weekend at the Sheraton Hotel. Crane had attended some of these symposia and had been struck at one by the concluding speech of the Boehringer representative, who had noted how enjoyable and useful the weekend had been, and that the company planned to fund more symposia 'provided that the sales are going well enough'. Fenoterol became popular among hospital-based specialists, and general practitioners followed their lead, with some encouragement from the company's sales reps who would visit regularly. Thus, the main factor influencing the prescribing of fenoterol in New Zealand appears to have been marketing. Certainly, none of the advertising claimed that the drug was in any way superior to the other beta agonists already on the market.[9]

A FORTE PREPARATION

Although fenoterol had not been marketed as being more potent than other beta agonists, Crane noticed that it had been formulated at 200 µg per puff, whereas salbutamol, the most commonly used beta agonist, was formulated at 100 µg per puff. The weight per puff is not a very reliable estimate of the strength, since different drugs can have different effects at the same weight. However, the larger weight per puff for fenoterol did raise questions as to whether the dose might be too high. Although the dose per puff was listed on the packaging, it seems that few New Zealand doctors had noticed it. Most doctors do not have time to read the labels and just assume, quite naturally, that the marketed dose is reasonable. As Crane later noted: 'One thing that really struck me was that very few people were aware that fenoterol was delivered in twice the dose of salbutamol'.[10]

It was later found that fenoterol may be twice as potent as salbutamol, weight for weight, meaning that 200 µg of fenoterol represents about four times the dose of 100 µg of salbutamol.[11] Thus, fenoterol was effectively a forte preparation, like isoprenaline forte. It is still not clear why fenoterol was marketed in such a high dose, or why fenoterol was not labelled and advertised at that time as being a high-dose preparation.

CARL BURGESS

Julian Crane discussed his concerns about fenoterol with Carl Burgess, a clinical pharmacologist in the Department of Medicine at the Wellington Medical School. Burgess had been in the department for a number of years, after getting his medical degree at the University of Cape Town and then working in the United Kingdom and South Africa. His work at the University of Southampton involved studies of the side-effects of drugs on the heart. The techniques involved are very specialised, and it was just by chance that Burgess happened to be in the right place at the right time.

Carl Burgess was not known for his subtlety, and his first reaction was sceptical. There had already been many studies of salbutamol, terbutaline and fenoterol. Although some showed a few more side-effects with fenoterol than with the other drugs, overall these studies did not suggest that there were important differences between them. If such differences existed, Burgess felt that 'surely someone would have noticed'. However, almost all of the studies had tested the drugs using just a few puffs. Burgess and Crane knew that their patients took much bigger doses when they were having a severe asthma attack. It was possible that fenoterol might behave differently when used in larger doses. Burgess finally agreed that a study would be worthwhile. He later explained: 'We went back to the literature to find out how much work had been done comparing these drugs in the way that people would use them during an attack, and there was virtually no work on it . . . we eventually found one paper. We felt that this was ridiculous, and it was time that someone had a look at it in detail'.[12]

Burgess and Crane were initially interested in a possible interaction between fenoterol and oral theophylline, since Doug Wilson had previously raised the possibility of a general interaction between beta agonists and oral theophylline. Crane was still employed as the Boehringer research fellow at this stage. Burgess later commented: 'We designed a trial, involving fenoterol, and sent it off to Boehringer for comment . . . The answer can back as an emphatic "no". They did not want us to study fenoterol in this way, but they wanted us to do it with salbutamol instead. . . So we did it with salbutamol and, much to Boehringer's dislike, we didn't find any problems with it . . . We published the work, and Julian finished his job as the Boehringer Ingelheim fellow. Then we decided to do a further study, involving fenoterol this time.'[13]

The study that Burgess and Crane designed included four treatments: isoprenaline, fenoterol, salbutamol and a placebo. Isoprenaline was chosen because it had been linked to the 1960s epidemics, and because it was known to be a non-selective beta agonist – in other words, it affects the beta-2 receptors in the lungs and thus relaxes the muscles in the airways, but it also affects the beta-1 receptors in the heart, producing strong side-effects. Salbutamol was chosen because it was the most commonly used beta agonist in the world and was considered to be the most selective – that is, it had far fewer side-effects on the beta-1 receptors in the heart. A comparison with a placebo (i.e. a harmless substance with no effect) was also made.

The study would involve eight healthy volunteers. Each person would come into the laboratory on four separate days, and would receive a different treatment on each day. Various measurements of respiratory and cardiovascular function (heart rate, blood pressure and other factors) would be taken before the participants had received any of the drug. They would then be given a few puffs and these factors would be measured again; they would be given a larger dose and the factors measured once more, and so on. Burgess and Crane felt that it would be unethical to give the participants the very large doses (of up to 50 puffs) that a patient might use in a severe attack. However, they felt that it was reasonable to give 18 puffs over a period of 45 minutes, particularly since the participants would not be hypoxic.

Having designed the study, Burgess and Crane sought some funding. They knew that the study would receive very little support, because the leading respiratory physicians considered that beta agonists were not dangerous. Even to raise the hypothesis meant questioning the work of the Asthma Task Force, which had declared that overuse of beta agonists was not the cause of the second New Zealand epidemic. However, at the time that the study was being planned, the Dean of the Wellington Medical School, Professor Ralph Johnson, was ousted because he had been trying to move the Wellington Medical School from being part of the University of Otago to being part of Victoria University, the main university in Wellington. He was replaced as dean by Tom O'Donnell, which meant he was no longer Burgess and Crane's head of department (as well as head of the Asthma Task Force). He was now one step removed and had much less influence and control over what they did. The new Professor of Medicine was Eru Pomare. He played a crucial role in the subsequent fenoterol controversy by doing nothing; that is, by resisting the pressure to 'rein in' the rogue researchers in his department who were causing so much trouble.

Fortunately, the study could be done with very little funding, since it was possible to use volunteers from within the medical school and the necessary equipment was already available. However, about $1500 was needed for special paper for the machines measuring the side-effects on the heart. Burgess and Crane applied to the Asthma Foundation, but the application was rejected by the Foundation's medical committee (chaired by Malcolm Sears, who was also a member of the Asthma Task Force) on the grounds that the committee thought that this was not an important area to pursue. Fortunately, Burgess and Crane had enough unspent funds from other grants to cover the costs of the materials, and they were able to go ahead with their study.

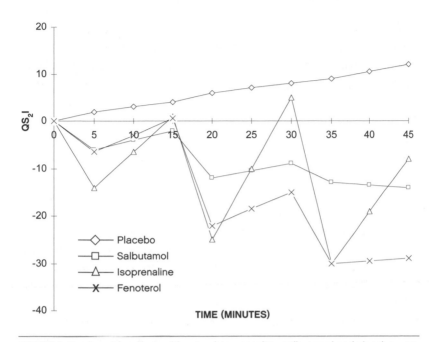

FIGURE 8: Cardiac side-effects of fenoterol, isoprenaline, salbutamol and placebo – as indicated by the effects on the QS_2 Index (QS_2I)

The participants were given increasing doses of 400, 600 and 800 µg from a metered dose inhaler at 15-minute intervals and heart rate, blood pressure and side-effects on the heart were measured. The findings astonished the researchers. Fenoterol was even worse than isoprenaline at increasing heart rate, blood pressure and other cardiovascular effects.[14] This can be seen in Figure 8, which shows the effects on the QS_2 Index, a measure of the workload and side-effects on the heart. The differences between the drugs were not evident after a few puffs (see the findings at 5, 10 and 15 minutes in Figure 8); in this situation salbutamol and fenoterol were similar to each other, and isoprenaline had much greater effects than salbutamol and fenoterol. However, the side-effects of fenoterol were longer-acting, and these accumulated as further doses were given; after a dozen or so puffs, fenoterol was showing much greater side-effects that salbutamol (see the findings at 35 minutes in Figure 8). Furthermore, these differences were occurring even though fenoterol was being used in the laboratory at 100 μg per puff (the same as salbutamol). Much greater differences might be expected with the marketed form of 200 μg per puff. More surprisingly, fenoterol was showing even greater side-effects than isoprenaline. As Burgess said later: 'It was the last thing we expected to find. We were stunned. Fenoterol was introduced to stop the effects on the heart that isoprenaline had, but the drug at its highest dose was worse than the one it was meant to replace.'[15]

The one related study in humans that had been done previously was by Tandon (a chest specialist in Western Australia), and published in the American respiratory journal *Chest* in 1980.[16] Tandon had compared the side-effects of fenoterol and salbutamol, and found that fenoterol produced a significantly greater increase in heart rate than salbutamol. Furthermore, some patients experienced cardiac arrhythmia (a precursor of a heart attack) with fenoterol. However, the study had produced little response from doctors. Most respiratory physicians were very reluctant to consider that such drugs could be dangerous, did not take the cardiac side-effects seriously and were unaware that these could be particularly dangerous if the patient was short of oxygen.

Many subsequent studies have confirmed that fenoterol has much greater side-effects on the heart than the other commonly used beta agonists, even at 100 μg per puff.[17] Furthermore, not only do the side-effects accumulate faster with fenoterol, but the maximum effect (with about 40–50 puffs) is much greater with fenoterol than with salbutamol. This seems to occur because fenoterol and isoprenaline are full agonists at the beta-1 receptors in the heart, while salbutamol and terbutaline are partial agonists at these receptors.[18] As a result, the side-effects of salbutamol and terbutaline on the heart reach a plateau fairly quickly, and beyond this point they do not increase even if further doses are taken. The side-effects of isoprenaline and fenoterol reach a plateau at a much higher level – if they plateau at all. Thus, the hazards of isoprenaline forte and fenoterol are not the result of their high dose alone, but also occur because they are fundamentally different drugs from the other beta agonists such as salbutamol and terbutaline.

These experiments certainly supported the hypothesis that fenoterol could have been involved in the epidemic of asthma deaths in some way. Not only had the epidemic started in the year that it was introduced, and closely paralleled the rise in fenoterol sales, but the drug was found to have similar side-effects to isoprenaline forte, which had caused the 1960s epidemics. This evidence was already stronger than that which had led to warnings being issued against the overuse of isoprenaline in the 1960s epidemics. Thus, there was a strong case to publicise the findings immediately, and to warn patients not to use fenoterol. However, Burgess and Crane were aware of the 1960s controversy, which had never been fully resolved because the definitive study had never been done. This would have involved a case-control study, comparing those who had died during the epidemic together with a control sample of people who had asthma but had not died, and examining which drugs the two groups had been prescribed. They knew the controversy that would occur once fenoterol had been linked to the new epidemic of deaths in New Zealand, and it was therefore important first to attempt to do a more definitive study.

The obvious first step was to look at the drug information from the Asthma Task Force's survey of asthma deaths.

RICHARD BEASLEY

Richard Beasley had had the same idea. Beasley had been a student at the Wellington Medical School, and had worked with O'Donnell, who arranged for him to spend two years on a research fellowship at the University of Southampton with Stephen Holgate, one of the world's leading asthma researchers. During his time in Southampton, Beasley had once sat next to Bono and the rest of the rock band U2 on a plane flight and didn't even know who they were – he had pointed out to them that the *Time* magazine he was reading had a band on the cover which was also called U2. Knowledge of the wider world was not his strong point, but he did know a lot about asthma.

Beasley had become fascinated with the strange epidemic of deaths in New Zealand. He was particularly interested in a paper

55

by the Asthma Task Force that was published in the *British Medical Journal* in February 1987.[19] The paper analysed 75 deaths in asthma patients who had been using home nebulisers; this issue had been considered because Grant had suggested that the misuse of home nebulisers might account for the epidemic in New Zealand. It was the only Task Force publication that mentioned which beta agonists the asthma patients had been using. This information had not been included in the original version of the paper, but had been added at the recommendation of a reviewer. In other words, it was there only because the *British Medical Journal* had insisted on this as a condition for publishing the paper. It showed that about half of the patients who died had been using fenoterol in their nebulisers, and about half had been using salbutamol. This contrasted with national sales figures, which showed that fenoterol accounted only for about a quarter of all sales of nebuliser units, with salbutamol accounting for most of the other three-quarters.

Thus, the study showed that the use of fenoterol by nebuliser was strongly associated with the asthma deaths. However, this finding was discounted in the published paper in two ways. First, the authors noted that the proportion of asthma patients using fenoterol was apparently much lower if attention was focused on the 10 minutes before death. Secondly, they argued that 'fenoterol was marketed after salbumatol, and patients with more severe asthma may have been transferred to the newer agent'.[20] In fact, the survey had not included any specific questions on the drugs used in the 24 hours leading up to death, and the information on use was apparently based on anecdotal reports from relatives, for those patients for whom a relative was present at the time of death. Furthermore, the maximum side-effects of such drugs usually occur after more than 10 minutes (just as a person's blood alcohol takes about 10–20 minutes to reach its maximum level after they have had a drink), and information on drug use within 10 minutes of death was probably not relevant. In addition, the survey involved deaths in 1981–83, which was five to seven years after fenoterol was marketed, and two to four years after it had gained a stable market share of 25–30 per cent. Any initial tendency for the 'new' drug to be tried out first in people with severe asthma would have worn off long before then, particularly since fenoterol had not been marketed as being any different from the other beta agonists.

The Asthma Task Force apparently had no intention of investigating the fenoterol findings any further, but Beasley found them fascinating. The obvious next step was to look at the data for all the deaths in the survey, not just for those who had been using nebulisers. On his way back to New Zealand in mid-1987, Beasley attended a conference in Australia, where Malcolm Sears presented some of the findings from the task force survey. Beasley asked him, from the audience, about the nebuliser data and whether the Asthma Task Force had examined the inhaler data. Sears replied that they had not looked at it and had no plans to. Beasley later commented: 'I was surprised, because that was the logical thing to look at next. I couldn't believe that, having found that, the Task Force was not pursuing it'.[21] Beasley asked Sears, in the public meeting, whether he could have a look at the relevant data, and Sears agreed.

On his return to New Zealand, Beasley was surprised to find that Julian Crane and Carl Burgess had been working on exactly the same idea:

> When I came back to Wellington I met Julian in the corridor.
> I didn't know him very well, but we talked about the work that
> we had been doing and planned to do . . . I mentioned that I was
> interested in the fenoterol findings from the nebuliser study
> . . . and he said that he had been interested in it for a long time,
> and he then told me about the Wilson paper, the pattern of
> relationship of the drug sales to the deaths, and the work that
> they were starting to do, looking at the effects of fenoterol on
> the heart. Putting it all together it was clear that, although each
> piece of evidence on its own didn't amount to much at all, when
> you considered it all together there seemed to be a number of
> coincidences that added up.[22]

They agreed that the logical next step was to look at the asthma drugs prescribed to the patients in the Asthma Task Force survey, and to compare this information with that from a sample of people with asthma who had not died. Beasley, Burgess and Crane agreed to join forces.

I GET INVOLVED

There's a joke about epidemiology being the art of turning death rates into Frequent Flyer miles, and there's some truth in that – if you are an epidemiologist, as I am, and do studies all around the world, as I do, you get to travel a lot. At least once a year the airlines lose my luggage, and each time I buy some new clothes on the insurance – something I count on doing so that I can get some reasonable clothes for work. So it happened that in early 1988 I was at an airport somewhere in Europe, my luggage had been lost for more than a day, I was tired and needed a shower and a good hotel bed. I was standing in the queue of the British Airways lost luggage counter. When I finally made it to the front, the woman behind the counter said,

'I hate this job, I've been working here since 6 am and my husband is waiting outside and I want to go home.'

'Well,' I replied, 'I'm a day late getting home to New Zealand and you've lost my luggage.'

'It seems that we're both having a bad day. The person in the queue before you told me, "I've come from Manchester and you've lost my luggage." I told her we didn't have any flights from Manchester. She looked confused and said, "Sorry, I've come from London". After 20 minutes of searching the passenger lists I worked out that she'd come from Amsterdam on KLM.'

By this time I wasn't sure where I was, but at least I knew where I had come from, and where I was going. I was on my way home from a meeting in Europe. What I didn't know was that when I made it back to Wellington my life was about to change. Beasley, Burgess and Crane had by now realised that what they were doing was actually called 'epidemiology', and I was the only epidemiologist in the Wellington Medical School.

My own entry into epidemiology had been rather unusual. The field involves a mixture of statistics, medicine and social science, and most epidemiologists have a degree in one of these fields and learn about the others as they train in epidemiology. In European countries, epidemiology has until recently been done mainly by medical doctors and there has been some pressure to keep out people with other skills. However, in other countries, particularly the United States, the public health 'medical mafia' is not so strong and epidemiologists have a broader range of backgrounds.

In my own case, I had done an honours degree in pure mathematics before getting interested in health issues. As part of my degree, I did a course in epidemiology, spending a couple of months learning how to predict epidemics. When an epidemic breaks out, its natural course can be predicted from information on the population at risk, the number of new cases every day, and so on. Almost all epidemics peak naturally, and the rate of new cases then starts to fall, even if nothing is done. At the end of the course, the lecturer told us:

> There is only one known use for these techniques. If you work for the World Health Organisation, as I did for seven years, and you get a call saying that there has been an outbreak of infectious disease somewhere – maybe an outbreak of cholera in Ghana – then you gather this information and plot the epidemic curve. Once you have worked out that the epidemic will peak naturally on the 17th, you send a fax saying that you will arrive on the 17th, and after you arrive the epidemic starts to go away.

Partly as a result of this inspiring teaching, as well as flatting with some medical students, I became interested in health research and unsuccessfully applied to be a medical student at Otago Medical School.* I then lost interest in the university and decided to get some experience of the 'real world' by working in labouring jobs and then as a bus driver for a couple of years. This was too much like hard work, so I tried to get a job as an orderly or a

---

* I was turned down on the basis that I was not of a high enough academic standard because I had taken five years to get my BSc. This was the result mainly of a long undiagnosed illness (brucellosis), but also of the usual 1970s student activities, including an extended period as technical editor for *Salient*, the Victoria University student newspaper.

cleaner at Wellington Hospital. There was nothing available, but I mentioned that I had an honours degree in mathematics. I found myself with a job as a biostatistical consultant to the Wellington Medical School, an ugly concrete building behind Wellington Hospital with all the ambience of a parking building the morning after an illegal rave party.

Straight away I became fascinated with epidemiology, and wanted to start doing my own studies rather than doing the statistics for other people's studies. I managed to complete a doctoral degree in epidemiology part-time before spending two years as a post-doctoral fellow in the United States at the University of North Carolina at Chapel Hill, funded by the New Zealand Medical Research Council.

I returned to New Zealand in September 1987, and it was about six months later that Richard Beasley came to see me, and started talking about fenoterol and the epidemic of asthma deaths. Although I knew nothing about asthma, I had heard about the epidemic of deaths, because there had been a lot of publicity about it during the 1980s. The problem sounded interesting and, as there was no one else that Beasley, Burgess and Crane could go to for help, I agreed to join the group. Thus, the Wellington Asthma Research Group was formed. However, I thought it very unlikely that they could have discovered the cause of the epidemic, since so many well-known New Zealand researchers had studied it for many years without success.

In some ways ours was a strange and unlikely group, as the four members each had very different backgrounds, interests and personalities – all four basic personality types were represented. The complementary personalities were important in the battles that were to follow. Even more important were the complementary skills. There was no other asthma research group in the world which had an epidemiologist (myself), a clinical epidemiologist (Crane), a respiratory physician (Beasley) and a pharmacologist (Burgess). Thus, the group was unique in the field of asthma research, and we had a uniquely difficult and interesting problem to tackle.

THE ASTHMA TASK FORCE DATA

The first job for Crane and Beasley was to look at the data from the Asthma Task Force survey, which were held by Malcolm Sears in Dunedin. They had planned to fly down to Dunedin for one day to look at the data, but Crane was refused permission to leave the medical school for a day by Tom O'Donnell, who was still formally his supervisor as well as being dean of the school. They therefore decided to travel on a Saturday, but on the Saturday morning Crane received a call from O'Donnell asking whether he was sure that he wanted to make the trip. Soon afterwards he received a further call from O'Donnell asking if he had permission from his head of department, Eru Pomare, to go. Crane ignored this and went ahead with the trip.

On their return from Dunedin, Beasley spelled out their reasons for making the trip in a letter to O'Donnell:

> In association with Julian Crane and Carl Burgess, we are reviewing the data from the Asthma Mortality Survey to determine whether excess beta agonist use can be implicated in the increase in mortality from asthma that has been observed in the last decade. In particular we hope to determine whether the newer beta-2 agonists introduced at the time of the initial increase in mortality are implicated in asthma deaths. As I discussed with you, our most recent work investigating the cardiovascular and metabolic side effects of beta-2 agonists suggests that there may be significant problems associated with their use.[1]

In a subsequent letter to O'Donnell,[2] Beasley asked for information on time trends in the use of beta agonists in New Zealand, and cited a previous task force recommendation that priority should be given to the study of trends in the use of asthma medications and trends in asthma deaths. He also asked for Crane's airfare to Dunedin to be paid from a small grant that he had received from the medical school to help establish his asthma research programme, but this routine request was turned down.

During the next few months, Beasley and Crane made further visits to Dunedin, and eventually abstracted all of the key Asthma Task Force survey data. This took some time because it had to be done manually, and checked again manually. Beasley said later:

> When we went down to Dunedin, we had to extract the data on all 271 deaths and enter it onto a computer . . . it was a major bit of work. We used to do it in the weekends, we would fly down on a Friday, we would work until very late at night, one person calling out the data, and the other person tapping it in. It was a mammoth job, and took us a very long time. Each time we went down we asked Malcolm Sears whether he had some of the data on a computer tape already, because we couldn't believe that such a vast amount of data could have been analysed by hand. . . . He said no, that all the tables had been prepared by hand . . . [so] we had to make three or four trips, and we had to borrow a computer to use.[3]

The first issue which Beasley and Crane looked at was the overuse of beta agonists. This was important because of the Asthma Task Force's claim that there was evidence for overuse of medication for only 3 per cent of the asthma patients who had died.[4] This claim had been widely quoted both in New Zealand and overseas, and had been taken as strong evidence that beta agonists were not the cause of the epidemic of deaths. Beasley commented:

> The Asthma Task Force had made the point very strongly that only 9 of the 271 patients had overused beta agonists during the attack in which they died, but when you looked at the data that was clearly not the case . . . In fact they didn't ask what drugs were used in the fatal attack in any case, so they had difficulty making any statement. However, when we used a very crude

form of analysis from the histories that were given in response to other questions, it was very clear that the number was much higher than nine.[5]

Even though there was no relevant information for many of the deaths, Beasley and Crane still found more than 100 people (of the 271) for whom there was evidence of overuse of beta agonists in the last few hours before they died.

They also found that more than half of those who died had been prescribed fenoterol by their general practitioners; a similar estimate was obtained using the reports from relatives. This contrasted with national sales figures, which showed that fenoterol accounted for about 30 per cent of the sales of beta agonist inhalers.[6] These findings for inhalers were similar to those previously reported by the Asthma Task Force for nebulisers.[7] Once again, they indicated that the proportion of those who died who had been prescribed fenoterol was much greater than would have been expected on the basis of national sales.

MEETING WITH THE ASTHMA TASK FORCE

Beasley, Burgess and Crane were in a difficult position. They had discovered more evidence that fenoterol might be causing asthma deaths. When this was considered along with the laboratory and time-trend data, it suggested that fenoterol could be the cause of the second New Zealand epidemic, which was still continuing. In the 5- to 34-year age-group (for which the data were most accurate) there was at least one 'excess death' per month; there were probably many more among older people. As the epidemic had been running for nearly 13 years, this meant that there had been about 150 'excess deaths' from asthma in 5- to 34-year olds, and probably several times this number among older people. It seemed likely that fenoterol was responsible for some of these 'excess deaths', but the new information that supported this hypothesis was from the Asthma Task Force survey and had not been published.

Beasley wrote a series of letters to the Asthma Task Force outlining their concerns. The first, on 8 April 1988,[8] requested a meeting with the Task Force to present the analysis of evidence of overuse in the survey data, and also to present the experimental study findings. This was followed, on 13 April 1988,[9] by a summary of their findings. On the same date, Beasley wrote to Harry Rea (a core member of the Asthma Task Force) asking if we could examine the data from the Auckland case-control study which had been published but for which the drug findings had not been reported.[10] He apparently never received a reply, but Rod Jackson, a co-author of the Auckland study, eventually gave us the data to look at.

I had begun to work with the group by this stage, but I was only too happy to escape the meeting with the task force. I had had no involvement in the work that was to be discussed, and I knew that the meeting was likely to be unpleasant. Instead, I was due to attend a cancer research meeting in Lyon, at the International Agency for Research on Cancer (IARC), an agency of the World Health Organisation.

Whenever I got to Lyon, somehow I ended up in a British pub called the Albion, and this time I quickly found myself there having a beer and telling my problems to John Kaldor, an Australian, who had made the bizarre decision to give up a successful rock music career to become a biostatistician, and now worked at the IARC. I sought his advice because the agency had much experience of being pressured by big business. There were problems not only with the agency's research, but also with its monograph programme, which provided authoritative evaluations of the carcinogenicity of various chemicals. I had been involved in some of these monograph meetings and had seen the pressure that industry brought to bear to try to influence the 'independent' experts that the IARC had invited. Most of the time, the IARC and the invited experts resisted the industry pressure very well, but I could see how, in other circumstances, the outcomes of the

---

* At that time, the IARC was headed by Lorenzo Tomatis, a wonderful Italian cancer researcher. After retiring back to Italy in 1993, he wrote a 'novel' called *Il Fuoriscito* (The Exile) about a cancer researcher who becomes head of an international cancer research agency, but becomes disillusioned because of interference by commercial interests and retires back to Italy.

evaluations could have been very different.* So Kaldor, and the other epidemiologists at the IARC, had had many problems with doing research which was unpopular with various companies. Nobody, however, had experienced quite the mess that I was becoming stuck in.

I was beginning wonder what I had got myself into. However, on these visits to Lyon I also often detoured to stay with friends in Paris, and it was on one such visit that I met someone who seemed like an excellent role model if my epidemiology career was to be ruined by fenoterol. My friends in Paris had been left-wing activists at university in Italy in the 1970s; as a result they could not get a government job in Italy, which meant that they could not work in an Italian university. Their friend Paolo was in the same situation, but stayed in Italy nonetheless. He was an extremely good poker player, and would be bankrolled by rich patrons for high-stakes poker games in the evenings. He earned enough to live on, and to write books on philosophy during the day.

On my way home I had to attend another cancer research meeting at Keystone, Colorado (in the Rocky Mountains). This involved one of the most bizarre flights that I had taken. It was from Frankfurt to Dallas and the passengers were 90 per cent German, but for some reason American Airlines chose to show a documentary about Kurt Waldheim, with long scenes of Nazi troops marching across Europe. As the plane was nearing Dallas, one of the stewardesses singled me out among 300 passengers and offered me half a dozen half-drunk bottles of wine from first class. That evening, the conference dinner had one bottle of wine for each table of 10 people (a common occurrence in the US), but I was able to keep our table supplied with half-drunk bottles from my room – though explaining why I had them was more difficult. It was only when I began to head home that I remembered about fenoterol, and the meeting that had been held with the Asthma Task Force on 5 May.

When I got back I found everyone angry and depressed. The meeting had not gone well. Crane and the others were allowed only a brief period during the lunch-break to make their presentation. Crane later commented:

From the outset, it was clear that they regarded this as a real nuisance. Here was a handful of idiots coming with a load of rubbish. I started by going over the isoprenaline story. At every juncture, I was interrupted by commments such as 'that's absolute rubbish, you can't say that', and so on . . . the whole meeting went on like that, and by the end people were shouting. It was a very bitter meeting, and it set the tone for the rest of the controversy.[11]

Towards the end of the meeting, Crane, Burgess and Beasley had presented the evidence of overuse of beta agonists that was apparent in the Asthma Task Force survey data, and had outlined the contradictions in the task force's claim that overuse could have been a factor in at most 9 of the 271 deaths. Crane commented: 'It became clear at the meeting that the task force had not used any criteria for overuse, it was based on what they called a "gut feeling". This was surprising, because so much had hinged on that comment in their publications.'[12]

One point on which the Asthma Task Force was emphatic was that the information from relatives was unreliable, and so it dismissed the evidence of overuse in the reports from relatives. As a result, we decided to restrict all of our subsequent work to the information from family doctors and hospital notes, and not to use any information from relatives.

The meeting had been a disaster. As Burgess described it:

If we had been students we would have been expelled . . . We were very naive and we didn't realise what we were getting into. We felt that doctors are there to make people better, and if they can't make them better, then to make them comfortable, but that's not the way it worked out . . . We presented our data, and we said to them that we believed that there was a case to be answered, that we should look at the drugs, the individual beta agonists. The meeting erupted, they were furious . . . it was awful.[13]

CASE-CONTROL STUDIES

We needed to do a proper study to find out whether fenoterol

66

really was linked to the asthma deaths. A randomised trial of fenoterol and asthma deaths would have been unethical, because of the evidence we already had that fenoterol was dangerous. A prospective study (i.e. recruiting people now and following them into the future) would have taken years and been impractical, as it would have needed an enormous number of patients (because asthma deaths are so rare). The only feasible approach was a case-control study.

The first modern case-control study was published in 1926, and this method is now perhaps the most common form of epidemiological research.[14] However, the first case-control study of asthma deaths was not published until 1985; this was a North American study of deaths in children, but it did not consider the role of asthma drugs.[15] The second such study was the case-control study based on the Auckland cases in the task force survey of asthma deaths, but this was very small and the findings for individual asthma drugs had not been reported.[16] So no one had previously done a case-control study to look at the role of asthma drugs in asthma deaths.

The basic idea of doing a case-control study of asthma deaths was to study all of the deaths as well as a control sample chosen from the population which generated the deaths. For example, for a study of all asthma deaths in New Zealand in 1981–83, the population under study (the population at risk) would be everyone in New Zealand who had asthma at that time – it would not include people who did not have asthma, because they would not be at risk of dying of the disease. The cases would involve all asthma deaths in New Zealand in 1981–83, and the controls would be chosen from people who had asthma during the same period but who did not die.

The two groups would then be compared to find out what drugs they had been prescribed. If a particular drug (e.g. fenoterol) was causing deaths, then we would find that a much higher proportion of those who died (the cases) would have been prescribed the drug, compared with a sample of people with asthma who hadn't died (the controls). However, if a drug was not causing asthma deaths, then the percentage of cases (deaths) who had been prescribed the drug would be the same as (or even less than) the percentage of controls who had been prescribed the drug.

It was time to call on Rod Jackson in Auckland for help. Jackson had been a friend for many years. We had never discussed the epidemic, but I had heard him speak about it several times. I remembered when he had spoken at the 1982 Public Health Association Conference in Christchurch and a television reporter had followed him down from Auckland to hear his talk and try to get him to comment publicly. Thus, I knew that he had thought about the issue for many years, and that he was one of the few people who believed that the epidemic could have been due to asthma drugs.

I was astonished by Jackson's response when I called him to talk about fenoterol. He had never heard of the drug. This, more than anything else, brought home to me the fact that no one had seriously considered the possibility that there might be differences between beta agonists. The issue had never arisen during all the years that Beaglehole and Jackson had been involved in investigating the epidemic. The whole debate had been about a possible class effect of beta agonists, and there had never been any discussion of individual beta agonists.

Jackson was sceptical at first, as I had been, because he felt that the issue would have been raised before if fenoterol was different from the other beta agonists commonly used in New Zealand. However, he agreed to help with the study. I was delighted with his response. Scientists are often not responsive to new ideas and may take offence, as had happened with the Asthma Task Force, at any suggestion that they might have got it wrong. However, Jackson was immediately keen to help with the study.

DESIGNING THE CASE-CONTROL STUDY

Over the next few weeks we held a series of meetings in Wellington to design the case-control study. We already had the information on the deaths in the Asthma Task Force survey of deaths from August 1981 to July 1983. We decided to study the deaths in patients aged between 5 and 45 years, because the epidemic had been most severe among this age-group, and the classification of asthma deaths is less accurate in older people.[17]

The next problem was to choose a control group to compare with the cases. The population we were studying comprised people with asthma aged between 5 and 45 years who were living in New Zealand in 1981–83. One obvious option would have been to choose the controls at random from this population. However, a major problem with this option was that less than 5 per cent of people with asthma have very severe asthma, whereas more than half of those who died had very severe asthma. Of course, all those who died must have had a very severe attack at the time they died, but only about half of them had had chronically severe asthma in the months before they died. It was chronically severe asthma that was relevant here because it might affect the beta agonist that such patients were prescribed in the months before the final, fatal attack.

Thus, it was possible that prescribing practices were different for patients with chronically severe asthma and those with chronically mild asthma. It was therefore important to be able to 'stratify' on levels of severity; that is, to be able to look at the subgroup of patients with chronically severe asthma and to make comparisons between cases and controls within this subgroup. Such a comparison would have been very difficult if most of the people who died had chronically severe asthma, but very few of the controls did. It was therefore necessary to find a way of selecting a control group that had a high proportion of people with chronically severe asthma.[18]

The solution was provided when we looked at the Auckland case-control study. This had used two control groups: one was based on 'community controls' of asthma patients chosen from family doctor records (the first option discussed above); the other was based on 'hospital controls' chosen from people who had had a hospital admission for asthma around the same time that the case had died.

The Auckland study had found many differences between the asthma deaths and the community asthma controls. In particular, those who died had much more chronically severe asthma. About half of those who died had had a hospital admission for asthma during the previous 12 months, whereas very few of the controls had. Patients with a recent admission were 16 times more likely to die than patients who had not had a recent admission. Of course, this did not mean that the admission caused the death; rather, it meant that people with a hospital admission for asthma had chronically severe asthma and were therefore at high risk of death.

The findings were quite different when the deaths were compared with the hospital admission controls. Most markers of asthma severity (including an admission in the previous 12 months) were similar in the cases and controls. The authors concluded that 'asthmatic patients who are admitted to hospital with asthma and patients who die appear to come from a similar portion of the asthmatic population – that is, they have troublesome disease (admissions to hospital), are non-compliant, and use accident and emergency departments for treatment of acute attacks'.[19] This meant that selecting controls from people who had had a hospital admission for asthma but who had not died would provide a reasonable 'match' on chronic asthma severity. Thus, this approach would give us enough numbers to compare cases and controls in the subgroup of patients with very severe asthma.

We therefore decided to use hospital controls, as had been done in the Auckland study. Thus, if someone had died of asthma in Wellington in July 1982, we would go to the Wellington Hospital records and choose at random, from listings of hospital admissions, four patients in the same age-group who had been admitted to Wellington Hospital with an asthma attack in July 1982 but who did not die. If we could not find all four controls in the records for July 1982, we would continue searching in the records for June and August, and so on, until we had found the four controls. Then we would pull out their hospital files and copy the information we needed. This process would be repeated around the country until we had chosen controls for each of the deaths in the study.

The next problem was to find out what drugs had been prescribed to the cases and controls. We had no intention of trying to find out what drugs had actually been used, because the information was not available. The Asthma Task Force survey had not included any specific questions on drug use in the 24 hours before death, and most people had been alone when they died, or with relatives only, and we had already been told by the task force that the information from relatives was unreliable.

Thus, we decided to base the study on the regular prescribed medication. We knew that this approach would be criticised because some patients might not have used their regular prescribed medication during the fatal attack. However, this was not a major problem provided that such problems of classification did not occur systematically. In any case, almost all asthma patients will use their beta agonist inhalers when they have a severe attack, and we later found only one patient whose prescribed beta-agonist medication had been changed during the final attack. So the regular prescribed medication was usually the same as the medication used in the final attack.

A further reason for focusing on the regular prescribed medication was that beta agonists are prescribed for long-term use (either 'regularly' or 'as required' for relief of symptoms), rather than for a particular attack. Any randomised trial of fenoterol would have to be based on long-term use. It would be unethical and impractical to recruit patients having an acute attack and then randomise their medication; only their regular prescribed medication could be randomised in such a trial. It was desirable that the case-control study should attempt to follow the approach that would have been used in a randomised trial. Furthermore, we were still not sure whether any problems with fenoterol stemmed from regular use or from use in the acute attack, so we had no clear guidelines for how to classify 'drug use' even had we been able to do so. So it was simpler, and more valid, to consider only each patient's regular prescribed medication, rather than try to work out what they had actually used in their final attack.

We already knew the regular prescribed medication for those who had died, because this information had been collected in the Asthma Task Force survey. The problem was to find out the same information for the controls, at the time of their hospital admission. Ideally, we should have got this information from their family doctors, as they had been the source of information for the cases. However, as we were doing the study in 1988, it was not realistic to try to find the family doctor of someone who had had a hospital admission for asthma some time during 1981–83, and to find out what their regular prescribed medication had been at the time of the admission. Even had this been possible, the accuracy of the information would have questionable, since it would have been collected five to seven years after the event, whereas the information for the cases had been collected within three months of the death. The only option was to use the information recorded in the hospital notes at the time of admission.*

We knew from the start that having to collect the drug information from different sources for the cases and for the controls was going to be a significant problem. This was a major design flaw in the study – the only one! – and we knew we would be criticised for it. However, the only alternative would have been to ignore the Asthma Task Force data, and set up a completely new study of deaths occurring over the next few years. We would never have got the funding or the co-operation required for such a study, and it would have been unethical to take several years to do it while the epidemic of deaths continued.

So we decided that the only option was to compare the information on prescribed medication from the Asthma Task Force survey for the cases with the corresponding information from hospital notes for the controls. However, we built into the study several checks to try to assess whether this approach was causing any serious bias. These checks (described below) eventually showed that the information from family doctors and from hospital notes was of similar quality, and there was no evidence of any serious bias.

Thus, the study was to involve 117 asthma deaths in the 5- to 45-year age-group during 1981 to 1983, and 468 controls

---

* The regular prescribed medication is routinely recorded at the time of an asthma hospital admission in New Zealand.

(four per case) matched on age and date of admission. The drug information had already been obtained from the family doctors for the 117 deaths, and it would be obtained from the hospital admission notes for the 468 controls.

Although our method of choosing controls from hospital admissions would achieve a reasonable matching for chronic asthma severity, it was important to carry out further checks for bias by analysing the subgroups of cases and controls who had chronically severe asthma. It is important to emphasise that it was *chronic* asthma severity that was important here, not *acute* asthma severity. This issue confused many people. Of course, the final attack for the cases would have been more severe than the corresponding attack for the controls; after all, the cases had died and the controls had not. However, the study was based on the regular prescribed medication of the cases and controls. This would not be affected by the final attack, as that would have happened months after the prescription was written; it would be affected only by the chronic asthma severity of the cases and controls over the previous year.

Thus, we needed to have some way of identifying cases and controls with chronically severe asthma. The answer was provided by the Auckland case-control study, which had found that the best marker of asthma severity (when cases were compared with community controls) was a hospital admission in the previous 12 months. We decided to look also at prescription of oral corticosteroids, since these drugs are prescribed only for patients with chronically severe asthma.

This study design was unintentionally supported from an unexpected source. Doug Wilson had become the Medical Director of Boehringer Ingelheim (NZ) Ltd, and was intensely interested in our activities. We had one meeting with him in which no one said very much because no one wanted to give anything away. At one stage Wilson opened his briefcase and a dictaphone recorder fell out. He was quick to tell us that it was not turned on, and Crane responded, 'But ours is!' After the meeting, Wilson sent us an internal Boehringer Ingelheim memorandum in which he criticised Burgess and Crane's laboratory work, and the time-trend data, and argued correctly that this evidence alone did not prove that fenoterol was dangerous. The memo then outlined how a more definitive answer to the issue could be obtained:

> By comparing the exact drug medication of patients who died with hospital controls matched on the basis of severity (of past hospital admissions, of recent steroid use, and other factors) then any major disparity between the representation of [fenoterol], or other drugs, could be taken as positive evidence, and strong positive evidence, of a selective toxicity of the over-represented agent.[20]

This was exactly the study that we were about to do.

AN INITIAL SKIRMISH

While we were planning the study, Burgess and Crane had attended a meeting of the British Thoracic Society in June, where they had presented the findings of the first laboratory study (see Figure 8).[21] They had not been given time for an oral presentation, but had been assigned to a 'poster' session.* Only two minutes were available to present the findings to the audience, and the critics seemed to be ready for them. The chair of the meeting introduced them as 'Burgess and Maclean', in a snide reference to the British spies who had defected to the then Soviet Union in the 1950s. The brief discussion was dominated by respiratory physicians sympathetic to Boehringer, and Burgess and Crane had the chance to say only a few words. As result, the paper had no impact among the hundreds of papers presented at the same conference. It was not a good start, and it gave us a taste of the problems to come.

---

* In an oral session you get 10–15 minutes to present your findings to a conference hall with several hundred people in it. In a poster session you prepare a poster of your key findings which is displayed on a noticeboard. You talk about it to the people who come to look at it, or sometimes there is a discussion of all the posters in a small room with a handful of people, and you get a couple of minutes to summarise your study.

We knew that things were going to be even more difficult in New Zealand, because the asthma field was completely dominated by the Asthma Task Force, which was funded by the Medical Research Council (MRC), and all applications to the MRC for asthma research projects were referred to the Asthma Task Force for comment. Furthermore, Tom O'Donnell served on the MRC in his capacity as Dean of the Wellington Medical School. The Asthma Foundation's medical committee was chaired by Malcolm Sears, and was similarly dominated by respiratory physicians involved with the Task Force. Despite these potential obstacles, we felt that it was important to try to obtain funding for the study, and to ask the Asthma Task Force to work with us on it.

We arranged to meet with Tom O'Donnell and Peter Holst, from the Asthma Task Force, on 18 July 1988. We showed them our application for funding from the Asthma Foundation, and asked them to become involved in the study and to work on it with us. They rejected our invitation to work together, dismissed the study as being unscientific, and advised us not to apply for funding to any body which would ask them to review the application.

We were delighted. We had already known that getting funding was going to be difficult. If members of the Asthma Task Force had agreed to work with us on the study, they could have made it impossible to complete, or at least could have delayed it for several years by continually disputing the methods and demanding alternative approaches or analyses. However, we had had an obligation to ask them to work with us because of the previous work they had done on the issue. We had asked them to work with us and they had refused, which for us was the best possible outcome.

We persevered in applying for funding to the Asthma Foundation and the MRC, but without success. The Asthma Foundation application was declined outright, whereas the MRC application was approved in principle, but was not funded because it had been scored at a low level of priority. Most of the reviews were very positive, but others were negative. One review was particularly dismissive:

The issue of beta-agonists contributing to asthma deaths is an old chestnut . . . I do not agree with the authors' statement [about the role of isoprenaline in the 1960s epidemics] . . . the applicants constitute a good group of investigators and are well placed to make a significant contribution to our understanding of asthma. However, I feel that the findings of the studies outlined in this proposal are unlikely to make that significant contribution.[22]

However, the funding situation was not too serious. The main funding we had requested was to pay the salary for a research assistant for a few months. We could still do the study if all of us, including the research assistants employed on other projects, were prepared to work on it for free in evenings and weekends. Most other expenses could be absorbed into the costs of other projects. All that was needed was some minimal funding for travel, and this was charged to some grants that the other three members of the group held for other clinical studies which had been completed within budget. On principle, I have never accepted research funding from drug companies, but in this case I was pleased that the others had had some drug company funding for their clinical studies, and had some funds left over. The main grant used was an old Boehringer Ingelheim grant which still had some money left in it.

BOEHRINGER TAKES AN INTEREST

By now Boehringer had heard about the proposed study, and Richard Beasley was receiving a constant flow of letters from Doug Wilson in his new capacity as medical director for the company in New Zealand. A letter from him dated 20 July 1988 stated:

Increasingly I am becoming concerned at aspects of the fenoterol toxicity hypothesis . . . It is clearly vital to Boehringer commercially, to you as a scientist, to the medical profession and public that this issue is examined with scientific precision before each or any is damaged by imprecise facts and inadvertent rumour . . . the company does not wish to market a dangerous drug but nor does it wish to delete a good one . . . some aspects of the industry are quite willing to use rumour for commercial ends irrespective of the damage it may cause to standards of medical practice, to patients' confidence in current therapies, and to other individual companies.[23]

So rumours were circulating about the planned case-control study, even though the grant applications had supposedly been reviewed confidentially. It was becoming important to do the study while it was still possible.

COLLECTING THE DATA

Collecting the data was the easiest part of the study. It was mostly done by Jenny Grainger, a visiting research nurse from the United Kingdom funded by a British Rotary Fellowship, who enjoyed the chance to tour the country and visit hospitals in interesting places. At the time, studies like this did not necessarily need ethical approval, because they involved looking at medical records without actually contacting the patients.* In some areas approval was granted directly by the local medical superintendent, whereas in others the study was referred by the superintendent for formal approval by the local medical research ethics committee. Whichever path was followed, approval for the study was eventually granted by all but one of the relevant hospitals. The only hitch came in Dunedin. Approval for the study was initially granted by the medical superintendent of Dunedin Hospital, but Malcolm Sears became aware that the study was being done, and a complaint was made to the local ethics committee. Further access to the Dunedin hospital admission data was blocked for nearly a year, so Dunedin was the only area not included in the study.

THE RESULTS

The initial results of the study were not particularly startling. The numbers changed a little in response to subsequent suggestions

---

* This was in the days before the Cartwright inquiry led to a major overhaul of the system of ethics approval in New Zealand. The inquiry arose out of evidence of unethical research being conducted without informed consent by Professor Herbert Green at National Women's Hospital in Auckland. It is described further in the concluding chapter.

from the Asthma Task Force, so the final published figures are used here to avoid confusion.

An understanding of case-control studies is crucial to what follows, so a detailed explanation is required. Let us suppose that a study had been done of everyone with chronically severe asthma in New Zealand, and they had been followed over the two year period 1981–83. The numbers might have looked something like this:

TABLE 1: Hypothetical data from a study of deaths in asthmatics followed during 1981–83

	Prescribed fenoterol	Not prescribed fenoterol	Total
Died	60	57	117
Survived	188,940	278,943	467,883
Total	189,000	279,000	468,000
Risk of dying (per 100,000)	31.8	20.4	25.0
Relative risk	1.6		

In this hypothetical example, the risk of dying from asthma is about 31.8 per 100,000 in those prescribed fenoterol, and about 20.4 per 100,000 in those not prescribed fenoterol. The relative risk of dying from asthma is therefore about 1.6 times in those prescribed fenoterol compared with those not prescribed fenoterol. However, to do a study like this would be a huge task, because it would involve studying 468,000 people with asthma and finding out who was using fenoterol and who was not, and then seeing who subsequently died. A case-control study is intended to get the same answer in a much simpler way, by studying all of the asthmatics who died and a sample of the asthmatics who did not.

Suppose that we had studied all 117 deaths and a sample of controls chosen by selecting one out of every 1000 asthmatics who did not die. The results would look like this:

TABLE 2: Hypothetical data from a study of deaths in asthmatics followed during 1981–83, and a sample of those who did not die

	Prescribed fenoterol	Not prescribed fenoterol	Total
Deaths (cases)	60	57	117
Survivors (controls)	189	279	468

The risk of death in each group can no longer be calculated directly, but the ratios can be. In those prescribed fenoterol, the odds of dying or surviving are 60/189; in those not prescribed fenoterol, the odds of dying or surviving are 57/279. The ratio of these two numbers (60/189 divided by 57/279) gives the same relative risk as before (1.6). The reason for this is obvious: 60/189 divided by 57/279 must be the same as 60/189,000 divided by 57/279,000. It is the same ratio as before: the two 'denominators' have each been divided by 1000 as we have sampled only one out of every 1000 people who did not die. More commonly in case-control studies, the data are presented the other way around: the odds of being prescribed fenoterol is 60/57 in those who died (the cases), and 189/279 in the sample of those who did not die (the controls). The *odds ratio* is therefore the same as before – that is, a relative risk of 1.6. So a case-control study gives us the same answer as if we had studied the entire group of 468,000 asthmatics.

While that is a hypothetical example of a case-control study, it corresponds to what we actually found:

TABLE 3: Numbers of cases and controls prescribed fenoterol

| | CASES | | CONTROLS | | | |
	Yes	No	Yes	No	Odds ratio	95% CI
Prescribed fenoterol	60	57	189	279	1.6	1.0–2.3

For each of the deaths we chose four hospital admissions for asthma, and for each group we found out whether or not they had been prescribed fenoterol. Of the 117 deaths, 60 (51%) had been prescribed fenoterol; of the 468 controls, 189 (40%) had been prescribed fenoterol. This translated into a relative risk of 1.55, which has been rounded up to 1.6. So the death rate of patients prescribed fenoterol was 1.6 times that of patients who were not prescribed fenoterol. This is only an estimate, but the 95 per cent confidence interval (95% CI) gives a range of values in which it is likely that the real answer lies.

So was this definitive evidence showing that fenoterol was causing the epidemic of deaths? Richard Beasley thought so, and was ready to try to convince others that 'this is it', that we had found the cause of the epidemic. I wasn't convinced. While there was a greater risk in those prescribed fenoterol, it wasn't very strong; it was the sort of risk (1.6 times) that could be caused by some sort of bias. In particular, it was possible (although unlikely) that the elevated relative risk of 1.6 could have been biased if fenoterol was prescribed more often to patients with very severe asthma, who might die because they had severe asthma rather than because they had been prescribed fenoterol.

Furthermore, the elevated relative risk was only just 'statistically significant' – that is, it was on the borderline of the sort of increased risk that could happen by chance. This is shown by the 95 per cent confidence interval of 1.0–2.3. So it was possible (but unlikely) that the true relative risk was 1.0 – that is, no increased risk at all. If you toss a coin 10,000 times and get 6000 heads and 4000 tails, then you can be fairly certain that there is something wrong with the coin and that your 'risk' of getting a head is about 1.5 times your risk of getting a tail. If you toss it ten times and get six heads and four tails, you would not be so convinced, even though the relative risk is the same. What we had found was somewhere in between these two scenarios: we had found an increased risk strong enough to be probably not due to chance, but we couldn't be sure.

An even more important issue was whether there was any evidence that the association of fenoterol with asthma deaths was due to bias. The best way to check was to look specifically at patients with very severe asthma. The potential for bias in this group would be less in the overall data-set (which included everyone in the study) since there was less 'room' for variation in average severity between the cases and controls (just as there is less potential for bias due to age differences if we do a study in 30- to 34-year-olds than if we do a study in 5- to 34-year-olds). If the overall relative risk of 1.6 went down towards 1.0 when we compared the more severe asthmatics using fenoterol with those using other beta agonists, this would indicate that the overall elevated risk of 1.6 had been due to bias. If it did not go down, this would indicate that the elevated risk was not due to this bias, and was likely to be real.

TABLE 4: Numbers of cases and controls prescribed fenoterol, stratified by asthma severity (based on an asthma hospital admission during the previous 12 months)

Admission in previous 12 months	CASES Yes	No	CONTROLS Yes	No	Odds ratio	95% CI
Yes	34	18	76	87	2.2	1.1–4.1
No	26	39	113	92	1.1	0.7–2.0
Total	60	57	189	279	1.6	1.0–2.3

You might think that all this statistical work is boring, but there is one fantastic thing about doing the data analysis of a study like this – you can really discover something new, and you are the first to know about it. You run the analysis, and the numbers appear on your screen, or on your printer. Usually, you don't find very much, and you are the first to realise that you have done a huge amount of work and the findings of the study are not very interesting. But sometimes you find something amazing. This is what happened here.

When we looked at patients with a hospital admission in the previous 12 months (the best marker of asthma severity), we found, amazingly, that the relative risk for fenoterol actually increased to 2.2 (see Table 4). In other words, whereas in general the patients prescribed fenoterol had a death rate 1.6 times that of those who were not, within the group of patients with severe asthma, those patients prescribed fenoterol had a death rate 2.2 times that of those who were not.

TABLE 5: Numbers of cases and controls prescribed fenoterol in various subgroups of asthma severity

Subgroup	CASES Yes	No	CONTROLS Yes	No	Odds ratio	95% CI
Admission in past year	34	18	76	87	2.2	1.1–4.1
Oral steroids	26	7	38	66	6.5	2.7–15.3
Admission in past year + oral steroids	18	2	21	31	13.3	3.5–51.2

Further, when we looked at the patients prescribed oral corticosteroids (a subgroup with very severe asthma), we found that the relative risk for fenoterol was 6.5 (see Table 5).

The most spectacular results came when we looked at the patients who had both severity markers: a recent admission, and oral corticosteroids. There were 20 deaths in this group, and all but two had been prescribed fenoterol; the relative risk was 13.3 (see Table 5).

I had never seen anything like this before (if they could, most epidemiologists would kill to get data like this). Not only were the relative risks really large; but, more importantly, we would not have seen these patterns if the association of fenoterol with asthma deaths had been due to bias. Instead, the data showed that fenoterol seemed to be particularly dangerous in patients who already had very severe asthma. The reasons for this were unclear. The high risk for fenoterol in patients with very severe asthma could be due to an interaction between fenoterol and oral steroids, or it could just indicate that patients with more severe asthma were more likely to become hypoxic and to use large doses of their beta agonist. Whatever the explanation, these large relative risks, and the patterns that we saw in the most severe subgroups, meant that the fenoterol findings were very likely to be real.

As soon as I saw this table, I knew that the study would create a storm of controversy when it was published. But first we had to get it published.

CHAPTER 6

Getting the paper published

One could not be a successful scientist without realizing that, in contrast to the popular belief supported by newspapers and the mothers of scientists, a large number of scientists are not only narrow-minded and dull, but also just stupid. – JAMES WATSON[*][1]

AN INITIAL SKIRMISH

It was going to take a month or so to prepare the paper for publication. However, Crane was enrolled in the Diploma of Public Health course, the fenoterol case-control study had been approved as his project, and it was time for him to submit a report. He prepared a draft report and, as a courtesy, gave it to Tom O'Donnell, who kept it for the weekend and then returned it without comment. Two days later, Crane received a letter from O'Donnell asking him to reconsider his priorities. Crane's half-time work on the public health course had been approved and funded by the Medical Research Council. The other half of his salary was provided by the medical school for teaching work, and was coming up for renewal at the end of the year. Crane replied with a lengthy letter asking O'Donnell to document the reasons for his concerns, and the matter was quietly dropped. Crane went ahead and submitted the report on his research project for the diploma course.

---

[*] This comment by James Watson, the co-discoverer of the structure of DNA, was made about physicists, but could equally apply to epidemiologists. As a species, epidemiologists combine the *joie de vivre* of statisticians with the Calvinist puritanism that permeates public health. At one meeting in Helsinki, the organiser told me that without doubt epidemiologists were the most boring group that she had ever hosted a meeting for – we ranked below molecular biologists and accountants. An epidemiology conference dinner usually has the ambience of a debriefing meeting of pallbearers who have just lost the coffin.

The next job was to let the Asthma Task Force know about our findings. We met with Rod Jackson on 11 November to discuss the study findings and plan our approach. We decided to take turns to meet with the various members of the Task Force in Auckland, Dunedin and Wellington, to be followed up with a formal meeting with all of the task force. The first meetings did not go well – the reactions ranged from incredulous to hostile – but worse was to come.

We had asked to meet with the full Asthma Task Force at its next meeting on 6 December, but instead an informal meeting was arranged with only the core task force members. The other members, including the Medical Research Council representative, were not going to have the opportunity to hear our case. Worried about what would happen at the meeting, we invited the Head of the Department of Medicine at Wellington (Professor Eru Pomare) to attend, since Beasley, Burgess and Crane were all in his department. We also invited the other researchers in our group, who had helped with the study, to attend. This was motivated partly by a need for moral support and a desire for 'safety in numbers'. However, we also strongly believed that scientific debate should not happen behind closed doors and that our colleagues should have the opportunity to hear about the study and to comment on it. We also wanted an outside person to chair the meeting. Our first choice was Robert Beaglehole, who was unable to attend, so we asked Rod Jackson to chair it.

The meeting started with an argument about who was going to chair it (it eventually proceeded without a formal chairperson), and why our colleagues had been invited. Heated argument broke out as soon as we began to present our findings. The Asthma Task Force members did not believe our data, and argued that the study should be delayed for six months while the data were checked. Fenoterol was still causing at least one death per month, so we felt that it would be unethical to delay for that length of time, particularly since the data had already been exhaustively checked. We therefore intended to submit the study for publication as soon as possible.

Following the meeting, Malcolm Sears wrote to us on 8 December, complaining:

> The presence of new junior members of your research group . . .
> was totally inappropriate to that occasion, and was inhibitory
> of frank discussion of the study and the ensuing issues which
> were not explored as they should have been. While I have
> the greatest respect for Eru Pomare, I saw no reason for his
> presence. Your actions in inviting these persons without seeking
> our views, or even advising us that they had been invited, was
> most discourteous and potentially inflammatory in a sensitive
> situation . . . I would like to know your reasons for extending the
> invitations outside those directly involved. You may regard me as
> one of the 'old guard' but I assure you I have your interests very
> much in mind in urging you not to proceed to publication of
> your data without further full and frank discussion of the study,
> the data, the message and the implications.[2]

Harry Rea's letter, also dated 8 December, was more direct:

> Your study has serious implications and I therefore believe
> strongly that it is in everyone's interests that 'no stone is left
> unturned' prior to publication. I recommend the following –
> (a) you seek legal advice.
> (b) You allow Boehringer Ingelheim full access to the data as
> soon as possible.
> (c) You seek an independent review of the raw data and your
> interpretations by reputable unbiased international experts.
> (d) I emphasize that you have an obligation to show the
> Medical Research Council Asthma Task Force how you
> intend to use the data collected by them. Both we and the
> Medical Research Council would clearly wish to establish our
> legal position.
> (e) I am sure that you also have an obligation to your
> employing institutions – no doubt you should show them any
> material considered for publication, as they will have clear
> legal responsibilities also.[3]

Copies of the letter were sent to other members of the Asthma Task Force, Eru Pomare, and the Director of the Medical Research Council, Dr Jim Hodge. We had already discussed the matter with a lawyer.* We had been advised that we should ignore any legal threats and proceed to publish the study. We were also advised that we should not inform Boehringer Ingelheim of our findings until the data were in final form – that is, when they were submitted for publication, or preferably after the paper had been accepted for publication.

This was the start of a long correspondence with the Asthma Task Force over the next few months, covering a wide range of issues about the study design and the accuracy of the data, and various legal and political issues. I now wonder what any of us expected to achieve by it all. No one was likely to change their minds, and the increasingly angry and aggressive letters just made the situation worse. It seemed that no one wanted to give the other side 'the last word', so it was always necessary to write another reply, even if no one else was going to read it. It is easy to regard such exchanges as trivial – Henry Kissinger, who worked at Harvard University before becoming Secretary of State under Richard Nixon, is reputed to have said that 'disputes between academics are so vicious precisely because the stakes are so small'. Anyone who has worked in a university will recognise the general truth in this remark, but in this case the stakes were a little higher than usual.

NEWS SPREADS

The day after the 6 December meeting with the Asthma Task Force, the Director of the MRC, Jim Hodge, called Richard Beasley. Hodge referred to alleged problems with the study design and to claims that the study had been done without ethical approval. In a letter of 7 December, he argued:

> It is clearly a matter of great public health importance, both within New Zealand and internationally, to determine whether

---

* The costs were covered by the Medical Protection Society which one of the other members of the group belonged to. The rest of us subsequently joined and I became New Zealand's first non-medical member.

the use of fenoterol is associated with an enhanced risk of a fatal outcome in asthma treatment. There is a corresponding obligation on investigators to be as certain as possible that their data warrant the conclusions which they draw from them. In view of the far-reaching implications of your own study, I consider that you and your colleagues would be well advised to give serious consideration to accepting an independent audit of your data prior to publication . . . I wish to propose that the Council itself be seen as the sponsoring agency for a small (e.g. two or three member) review panel of the highest repute and integrity to conduct a fully independent review of your study data prior to their publication.[4]

We were particularly concerned about the proposal for an MRC review of the study, because we felt that the Asthma Task Force, which was an MRC committee, would have undue influence over the process. This concern was heightened when Hodge suggested that such a review could be conducted by Ann Woolcock, Professor of Respiratory Medicine at the University of Sydney. Although we had great respect for Woolcock, she was an old friend of the task force members, and had previously been in New Zealand as a Boehringer Ingelheim Visiting Fellow. Furthermore, although a very good scientist, she was not an epidemiologist. Our concerns were subsequently justified when she participated in a Boehringer Ingelheim consensus panel which criticised our study.

The company itself was also getting into the act. Soon after our meetings with individual members of the Asthma Task Force, Boehringer appeared to know a great deal about the study. On 5 December, Doug Wilson wrote to Beasley stating:

My parent company . . . requests the following of you:-
  1) All data accumulated by you should be made available to Boehringer Ingelheim at the earliest opportunity for evaluation.
  2) A panel of 3 internationally eminent respirologists/ epidemiologists, with unquestionable authority in the conduct and assessment of epidemiological studies involving asthma mortality will be involved, together with the Medical Research

Council of New Zealand, to clarify these extremely serious issues.

3) Boehringer Ingelheim will support any action suggested by this panel to help clarify the issue.[5]

The panel was to include the familiar name of Alvan Feinstein (Yale University), who had co-authored the review which claimed that the 1960s epidemic may not have occurred.

The letter was followed by a visit from representatives of the parent company in Germany, including Dr Victor Hartmann, Head of Central Department Medical Services. We declined to meet them on the ground that the report on the study was not yet in final form. Instead they met with the Otago University administration down in Dunedin, and after that Beasley had to see the Vice-Chancellor.

The situation was clearly getting out of control, and the prospect of collaboration between the MRC, the Asthma Task Force and the company to condemn the study was worrying. The only way to overcome the pressures from the MRC and the university was to go over their heads. The first move was made by Laurence Malcolm, head of the Department of Public Health at the Wellington Medical School. He wrote to O'Donnell on 9 December:

> I am writing to express my serious concern about the lack of action being taken with respect to the findings of Neil Pearce, Julian Crane, Richard Beasley and Carl Burgess regarding fenoterol in the treatment of asthma. Their case-control study appears to provide convincing evidence that the use of fenoterol is associated with substantially higher death rates than alternative drugs, especially in asthmatics with serious disease.
>
> In view of the potentially serious ethical consequences I strongly urge you to formally advise the Department of Health as soon as possible that you have become aware of evidence regarding the safety of fenoterol which would appear to require that it be withdrawn from use.[6]

Laurence Malcolm himself informed the Director-General of Health, Dr George Salmond, about the study, and he in turn informed the Minister of Health, David Caygill. The next day I was called by Salmond, and we arranged to meet with Department of Health officials two days later, on 14 December 1988.

THE DEPARTMENT OF HEALTH

When we arrived at the Department of Health, Boehringer had been there first. Wilson had met with Salmond the previous day, and suggested that we may have changed the protocol during the course of the study. This was completely untrue. Such an allegation is serious and can undermine the credibility of a study, as well as the credibility of the researchers. It is not unusual to make minor changes in the data collection methods as a study proceeds and the practicalities become more apparent, but it is important that the overall protocol is closely adhered to. To accuse researchers of having changed the protocol in any significant way is the same as accusing them of scientific fraud. We never established what aspect of the protocol had allegedly been changed, and Wilson eventually retracted the allegation.

The meeting at the Department of Health went well. Its representatives included Salmond, Warren Thompson (head of the Medicines and Benefits section), Phil White (a general practitioner temporarily working with the department) and Karen Poutasi (a public health specialist who eventually became director-general). I presented our data to them and finished by saying, 'This is the sort of data that you only find once in a lifetime.' They jokingly replied, 'We can arrange that.' They were as surprised by the findings as we had initially been, and obviously the data posed some major dilemmas for them. However, they seemed genuinely concerned about the situation, and how the department should respond to it, and wanted to hear what we had to say. We felt that for the first time we were talking to people who were responding logically to the evidence, rather than with a knee-jerk reaction in defence of Boehringer Ingelheim or their own reputations. Over the next few months, I discovered that first impressions can be wrong, but at least we got off to a good start.

In the next few days, Salmond consulted our group and the Asthma Task Force in order to set up a review panel. It was to be headed by Stephen Leeder, Professor of Community Health at the University of Sydney, and to include Robert Beaglehole and Dr Charlotte Paul, a lecturer in community health at Otago University. We accepted this proposal, and told Hodge that there was no need for the MRC to set up a review panel, as the Department of Health had already created one.

THE DEPARTMENT OF HEALTH REVIEW
The Department of Health review panel met on 20–21 December 1988 in Wellington. The first day involved separate meetings with our group and with members of the Asthma Task Force (O'Donnell, Holst, Sears; Rea was unable to attend). We all met jointly the next day. Although the review panel had some criticisms of the study design, and particularly about the collection of the drug information from different sources for the cases and controls, the meeting concluded that 'they would accept the study design as appropriate to the study and that . . . the findings of the study would be sufficient to justify public health action. The study had several significant difficulties but these were judged not to be of sufficient magnitude to render the findings irrelevant to public policy.'[7]

We thought that the objections of the Asthma Task Force had now been overcome, and that we could proceed to publish the paper. However, towards the end of the meeting we were told that the task force had questioned the accuracy of the data in the case-control study. Malcolm Sears presented an analysis of the data on nebulisers which seemed to be very different from our own results. The review panel was placed in a difficult situation, as Sears was suggesting that there were serious mistakes in our analysis. Although the review panel had approved the study design in principle, it could hardly recommend that the study be published immediately when it was being claimed that there were mistakes in the data. We were seeing Sear's analysis for the first time, and did not know how to respond to it, but we knew that our tables were correct, since we had checked the data many times. Crane later commented:

We were most surprised at the task force's claims, because we had been so careful at checking and double-checking the data, because of the importance of the results. We had checked the data almost to an obsessive degree. Thus, we knew that the task force was wrong, because we had already double-checked it . . . When they came out and claimed that they couldn't reproduce our findings, we were absolutely amazed, and had great difficulty accepting that the submission of the paper should be delayed.[8]

The Asthma Task Force members were insistent that the study should be delayed by six months while all the data were rechecked. We argued that a check of the data could be performed within two weeks. After a heated argument in which the meeting came close to collapse, it was agreed that the paper would be delayed for six weeks so that the data for all of the deaths and a sample of the controls could be rechecked jointly by us and by the task force.

We knew that this review would show the data to be accurate, but we were concerned that the process would never end. Once the data had been shown to be accurate, the task force could then return to the previous criticisms of the study design, or dispute the data on prescribed medication by using data from other sources, such as relatives, or raise entirely new criticisms. We therefore argued that the data review would be justified and feasible only if the task force accepted that the study design itself was valid. After heated discussion, the task force members finally agreed to this condition. As the minutes of the meeting recorded: 'Discussion occurred as to whether evidence of use of a beta agonist in the fatal attack was essential. IT WAS AGREED that it was not and that the case-control study was based on PRESCRIBED medication'.

After the meeting we finally got the chance to check the task force's new figures, and it became clear that they did not contradict our tables. Sears had different figures because he had included data on nebulisers used for other drugs (as well as beta agonists) along with information from relatives. He had also made a simple adding mistake. For this, the publication of the findings was going to be delayed once again. We were particularly angry about the use of data from relatives, because the task force had told us so emphatically six months earlier that this information was unreliable and that we should not use it.

Two days later was the last working day before Christmas, and we finished the year with a few beers in Beasley's office. We were depressed that, after all the effort and the strife, involved in doing the study, it had now been delayed because of claims about the accuracy of the data. It was easy to see that these claims could continue indefinitely, and that we would eventually have to try to publish the paper with or without the agreement of the task force. However, we first had to go over the data with the task force members once again, revisit all of the arguments, and make a final effort to gain their co-operation.

VALIDATION

The first task for the new year was to conduct the validation exercises with the task force. The job fell to Richard Beasley while the rest of us took our summer holidays. When I got back from two weeks travelling around the South Island, I found that he had done a fantastic job. He had flown to Nelson to collect the original data for the deaths from Malcolm Sears, who was on holiday there. He had then been through all of this data with Peter Holst, another task force member. There had been very few discrepancies, and those changes that Holst recommended had strengthened our findings. This was because we had been careful to avoid any suggestion of bias and, when the correct classification of the data had been unclear, we had taken the option that counted against our hypothesis. In particular, if we had not been completely certain that a patient who died had been prescribed fenoterol, we had assumed that they had not. When Holst checked the data, he 'corrected' some of these classifications so that we ended up with a slightly higher number of patients who were considered to have been prescribed fenoterol.

Beasley had also been through a sample of the controls in Dunedin with Sears, in Wellington with Holst, and in Auckland with Rea. In Wellington, an independent validation had also been performed by David Jones, a chest physician who was on the Medical Advisory Committee of the Asthma Foundation. In each instance, the agreement had been very good and very few discrepancies had been found. Jones had reported: 'I was slightly surprised at the ease with which drug information was obtainable from the notes. . . In comparing my findings with those of the original researcher, for bronchodilator treatment there was agreement in 51 out of 52 records'.[9]

Similarly, Rea had concluded: 'I have no concerns about the accuracy of the data extraction, nor about the importance of the work and the investigators' industry and sincerity . . . [I] agree . . . that this is the best data that will ever be available to try and answer this very important question.'[10] However, in his report Rea returned to his previous criticisms of the study design, which had already been resolved by the review panel. He also claimed that, for some of the controls, asthma had not been the reason for their hospital admission, even though they did have asthma. In fact, for all of the controls, asthma had been coded in the hospital records as the reason for their admission; for all patients for whom information was available, it was clear that the doctors who treated them had themselves considered that asthma was the main reason for the admission. Despite this, Rea's claim was often repeated in subsequent months – I heard it myself at many scientific meetings – and it became accepted as a major problem with the study.

Nevertheless, the validation had generally gone well, and showed that, although there were still disagreements about the general study design, there was very good agreement on the classification of the key drug data. On 27 January 1989, Beasley wrote to inform O'Donnell that the changes suggested by the task force had been accepted, and that the manuscript was about to be submitted for publication.[11]

DEFERRING A DECISION

While the validation was being done, Boehringer Ingelheim had continued lobbying the Department of Health, and had asked George Salmond to state whether any immediate action was necessary over fenoterol. Although the review panel had recommended

93

that public health action was justified, the department's Medicines Adverse Reactions Committee (MARC) had not yet considered the issue, and on 6 January 1989 Salmond replied:

> At this point I am satisfied that no pre-publication action is required by this office to in any way influence the availability or use of fenoterol in New Zealand. At present my intention is to wait for reaction to the published paper before contemplating further action if, indeed, any is required.[12]

This statement was subsequently quoted many times by Boehringer Ingelheim to claim that there was no reason to be concerned about the safety of fenoterol.

SUBMISSION FOR PUBLICATION

The paper was sent to the *Lancet* on 3 February 1989. As we put it in the fax machine, I said, 'That was easy – now comes the hard part'. This turned out to be truer than I had expected.

When we submitted the paper, we gave a copy to the Department of Health, because we felt, and had been advised by our lawyer, that we had a moral obligation to do so. We had said many times to the department that we were concerned that there might be interference in the normal process of scientific peer review and publication if the manuscript were given to Boehringer Ingelheim. However, we finally agreed that Salmond could give a copy to the company on the written condition that the company would make no attempt to interfere with the paper's publication. Accordingly, on 7 February 1989 a copy was sent to Doug Wilson. In the covering letter, Salmond stated:

> I am providing the company with a copy of the paper for information and to enable it, if it wishes, to make written submissions to the Medicines Adverse Reactions Committee . . . As explained on the telephone I am providing this copy of the draft paper for the company on the understanding that it will not seek to influence the normal process of publication.[13]

The same conditions were spelled out in a similar letter from Salmond to the *Lancet*.[14]

A SURPRISE IN OXFORD

As well as being famous for his own research, Sir Richard Doll is a legend among epidemiologists for his perceptive advice on other people's research.* In particular, Doll had been helpful with Rod Jackson's earlier research on the second New Zealand epidemic.[15] At Jackson's request, he had reviewed an earlier draft of our paper, and (with his permission) we had acknowledged his help in the manuscript.

Soon after the paper was submitted I had to make a trip to Europe for a cancer research meeting at the IARC in Lyon, and I decided to make a detour through Oxford to see Sir Richard Doll. I had been going through a period of being very nervous about flying, which had started when I had an incredibly rough landing at Wellington Airport a year before. It had been made worse by my general stress and nervousness about the fenoterol work, but I had to keep flying to do the work. When I got to Oxford, I opened the newspaper to find a picture of a United Airlines 747 with half the roof missing. It had taken off from Honolulu and a cargo door had accidentally opened, blowing the side off the plane and sucking several passengers out. I had come on the same flight, and probably the same plane, just two days earlier.

When I arrived at his office, he had a fax from Boehringer lying on his desk. The company had seen the acknowledgement to him, and had faxed him a copy of its internal review of the manuscript. We photocopied the lengthy document on our way to lunch in the refined English surroundings of the dining room in the old

---

* For example, some colleagues in Barcelona had consulted him about a series of outbreaks of severe asthma attacks in that city, which they had studied for years without finding out the cause. Doll agreed to meet them for an hour on a train from Oxford to London. He listened to their story and, when the train was nearing London, he suggested that they find out what was happening at the port. They went back to Barcelona and discovered that the epidemics were linked to the unloading of soyabeans into unfiltered silos at the port. (J.M. Anto and J. Sunyer, 'Epidemiologic Studies of Asthma Epidemics in Barcelona'. *Chest* 98, 1990, pp.S185–S190). The silos were covered over and the problem went away.

observatory at Green College, and I started to read it over lunch. It was dated 8 February 1989 – that is, one day after Boehringer had supposedly received the manuscript for the first time. It stated:

> This commentary has been developed after discussions with Mr AG Dick, Mr HG Ngan of BINZL, and telephone communication with Drs H Rea and J Kolbe of Green Lane Hospital, Dr W Wardell and Dr J Wahl of BIPI, Professor Michael Weintraub of the University of Rochester New York, and Professor Malcolm Sears of the University of Otago.[16]

So my first shock was to find that the company had consulted with two Asthma Task Force members. The Boehringer report then stated:

> There may be no formal protocol for the study. Information on this defect was provided by two Task Force members . . . It appears that the protocol as such was developed during the course of the study as information came to hand . . . the authors state that the primary focus of this case-control study was on the possible role of self-administered inhaled fenoterol in asthma mortality. As there appears to be no formal protocol to begin with this statement is difficult to sustain.

This statement was, of course, false, since the protocol had been included in the grant application that was rejected by the Asthma Foundation and the Medical Research Council.

The document then discussed issues which had been raised during the formal meetings with the task force, including the questionable analyses of information from relatives. It stated, for example:

> The group . . . commented on the use of the home nebulizer study of Malcolm Sears which they believe supported their argument by stating that half of the patients with fatal asthma were given nebulized fenoterol . . . Professor Sears in his careful and detailed analysis of therapy actually used, rather than

prescribed, firmly indicated that there was no overrepresentation of fenoterol . . . The authors were provided with this information but chose to ignore it.

The Boehringer review also included a memorandum from Dr Joanna Haas, an epidemiologist with the company in Germany, who argued that 'This study should be attacked at its roots, i.e. the study design'.[17]

I felt that the Boehringer document was personally damaging, as well as damaging to the study, since the false allegations that the study had been done without a protocol, which followed on from the previous allegations that the protocol had been changed, were close to accusing us of cheating, and would prejudice any reviewer's assessment of the study. What was particularly disturbing was the material from the Asthma Task Force, not only because it included false allegations, but also because it appeared to contain information from our discussions with the task force and the Department of Health review panel which we had believed were confidential.

We also later learned that the task force had written to the *Lancet*, while the paper was still under review, requesting permission to publish a reply if it was accepted for publication.

NEWS IN LYON

I arrived in Lyon the next day, and after a stop at the Albion faxed the Boehringer document back to Wellington. During the first session of the IARC meeting, a fax from the *Lancet*, dated 20 February, was delivered to my desk. It stated:

We will publish your paper on fenoterol and asthma mortality. We have reports from two referees who raise small points . . . I understand that the manufacturers have sought advice from their own referee, a copy of your paper having been sent to them by Dr Salmond. If you decide to change your paper in the light of those comments would you please send us a copy of the other referee's report – otherwise our final view of the paper will be confused.[18]

97

I was pleased that the paper had been accepted, but I was confused by the reference to the Boehringer review. I wondered what it was, and how the *Lancet* knew about it.

THE BOEHRINGER CONSULTANTS SWING INTO ACTION

I returned to find that the others were feeling more nervous than they had when I left. As soon as I arrived, they came out to my house and we had a beer in the back garden. They were concerned because the first reviews commissioned by Boehringer had begun to arrive at the Department of Health, which had passed them on to the Medicines Adverse Reactions Committee.

The first two reviews were in the form of letters from the two groups that had previously worked for Boehringer Ingelheim on the issues of orciprenaline in the United States, as described in Chapter 2.[19] Feinstein's review was fairly direct:

> I shall be very surprised if the work is accepted by the *Lancet*
> ... The editors of the *Lancet*, however, sometimes do not seek
> a careful external scientific peer review; and many editors
> nowadays are attracted by potential publicity for published
> reports. If the work is accepted and published, the devastating
> criticism and subsequent embarrassment (for authors, editors,
> etc) will occur afterward.[20]

The other review was by Lanes and Walker of Epidemiology Resources Inc. They gave more constructive criticisms, some of which were useful although they dealt with trivial issues. The review also outlined their approach:

> Our approach was to review the study, to identify potentially
> important sources of bias ... The next step would be to
> evaluate the importance of each source of error ... Exposure
> classification is a concern because subjects were classified
> according to drugs that were prescribed but not necessarily
> used ... Controls were matched to cases by region; however,
> this variable was not considered a potential confounding
> variable.[21]

What they did not mention was that these two 'sources of error', as with many other 'problems' listed, would tend to produce false negative rather than false positive results, and therefore could not account for the positive findings of the study.

The other reviews were little better. They were strangely emotional in their defence of the drug and of Boehringer Ingelheim, and in their criticisms of the study and of our group for doing it. Another feature was their lack of consistency. For example, Sonia Buist (Oregon Health Sciences University) stated that 'The first and probably most important problem [with the study] is the absence of an initial hypothesis',[22] and referred to the alleged lack of a formal protocol. A review by Anthony Rebuck (University of Toronto) stated that 'The authors [of the study] have developed a hypothesis which they have pursued relentlessly'.[23] Walter Spitzer (McGill University) and Ralph Horwitz (a colleague of Feinstein at Yale University) also referred to the alleged lack of a protocol and criticised the study for using 'historical information which was as much as four or five years old at the time of data collection'.[24] In fact, the data had all been recorded in the hospital and family doctor notes at the time of the relevant events (death or hospital admission), and had merely been copied when we did the study five years later.

Some reviews were not so extreme, including those by Howell and Waters (University of Southampton)[25] and Peter Burney (St Thomas's, London).[26] However, the only review which was moderate and comparable to those commissioned by independent bodies (such as the New Zealand Department of Health) was that by Sackett, Browman and Marrett at McMaster University in Hamilton, Ontario.[27] Boehringer was unhappy with this review, and apparently tried to persuade the authors to change it, as is alluded to in a further letter from Browman to Hartmann (Boehringer Ingelheim, Germany): 'I reviewed our conversation and the additional documentation provided to us with Dave Sackett. As I indicated to you over the telephone, we do not believe that the new information provided warrants any change in our original submission.'[28]

The overall weight of the Boehringer Ingelheim reviews was

overwhelmingly negative, and the evidence against fenoterol was ignored or dismissed. The criticisms made were very unlikely to explain our findings, and many of them were absurd to anyone with training in epidemiology. However, they were also highly technical, and could appear convincing to people without training in epidemiology. At best, they could convince naive readers that the experts could not agree; at worst, they could convince them that the study was fatally flawed.

I did not take the reviews seriously because I knew that we had done a good study, that I was reasonably competent, and that I had a good reputation internationally among epidemiologists. However, all of this doesn't count for much in New Zealand, where most people believe that someone from New York or London must be more of an expert than someone from Wellington. It also doesn't count for much among doctors, most of whom know little about epidemiology.

This was really brought home to me in our 'beer garden meeting' when I found that even the other members of the group had been affected by the criticisms and had begun to wonder whether we should withdraw the paper. Even they did not understand the methods of epidemiology well enough to realise that the criticisms were mostly not worth taking seriously. Crane was particularly concerned about the Spitzer review because he had confused Spitzer with Speizer, who had done the original studies of the 1960s epidemic,[29] remarking that 'Even the guy who studied the 1960s epidemics doesn't seem to believe us.' Once I explained that Spitzer and Speizer were two different people, things calmed down, and we decided to go ahead with publishing the paper.

MARC SUBMISSIONS

Armed with these reviews, Boehringer Ingelheim now set to work to lobby the Department of Health. Unfortunately, the department did not have an epidemiologist on its staff, nor did its Medicines Adverse Reactions Committee, so neither body had the expertise to see the obvious flaws in the reviews.

Before the company started lobbying, it seemed as though the department was going to restrict the availability of fenoterol, or even to withdraw it completely. MARC was scheduled to meet on 24 February 1989, and it was expected that the decision would be taken at that time. However, the department advised the Minister of Health that they could be sued if the company did not have enough time to consider the evidence and make a submission. The meeting was therefore postponed until 22 March. In the meantime, the Boehringer reviews continued to arrive at the department. The Minister of Health was now Helen Clark, as David Caygill had moved on to replace Roger Douglas as Minister of Finance as a result of upheavals in the Lange Labour Government.*

Boehringer Ingelheim also made its own lengthy submissions to MARC.[30] These included a further critique of the study from Lanes and Walker, and a study of the use of fenoterol in six dogs which was claimed to refute our laboratory studies in humans. The submissions also included graphs which claimed to demonstrate a lack of association between fenoterol sales and asthma deaths in various countries. For example, a graph was presented of fenoterol sales and asthma deaths in New Zealand which showed fenoterol sales rapidly rising during the 1980s while the death rate rapidly declined; however, two crucial points on the graph were in the wrong place, and a corrected graph showed a strong correlation between fenoterol market share and asthma deaths. Another graph which claimed to show no correlation between fenoterol sales and asthma deaths in West Germany was drawn on the wrong scale, and when I produced a corrected graph it once again showed a strong correlation between fenoterol sales and asthma deaths.[31]

The Asthma Task Force also made a lengthy submission to the Department of Health.[32] This repeated the criticisms of the study design which had already been discussed and resolved at the review panel meeting in December 1988, and argued, without

---

* This change was to prove crucial, as Clark was a more assertive Minister of Health. It probably also helped that she was married to Peter Davis, a lecturer in Robert Beaglehole's Department of Community Health at the Auckland Medical School. Thus, she was more sympathetic to public health and more willing to take epidemiology seriously, and she certainly had plenty of good advice readily available.

providing any evidence, that our findings were 'fully explained by the greater use of fenoterol in patients with more severe asthma'. The submission repeated the claim that some of the controls did not have asthma (this was subsequently quoted by the company as having been stated in a report 'obtained from the Department of Health', but in fact the report was a submission from the Asthma Task Force to the Department of Health). The task force submission also included a further review from Michael Leibowitz at the University of Arizona,[33] and a covering letter from Malcolm Sears stating that Professor Leibowitz's 'frank comments would be that the investigators are amateurs at epidemiology and do not seem to know a great deal about asthma'.[34]

THE *LANCET* GETS COLD FEET

By now things were beginning to get us down. We had known that the study would be criticised, but we had not expected the level of personal criticism we were getting. News about the study had spread widely, and we regularly received calls from newspaper reporters who wanted to run a story on it. Each time we had managed to persuade them that it would be irresponsible to publish the story until the paper itself had been published. It was important for doctors and policy-makers to have the full published report available rather than having to rely on brief and possibly inaccurate newspaper reports. It was getting more difficult to hold the reporters back, however, and in return for their 'silence' we had to promise an exclusive interview when the study was finally published. By this time we had promised several exclusive interviews, so we couldn't hold out much longer.

Some people criticised us for not going public with the findings straight away. After all, people were dying and it was going to take some time to get the paper published, so why not go straight to the newspapers? However, if you do that, then no good scientific journal will publish the paper. There is a good reason for this policy, as it is important that any report is fully peer-reviewed (i.e. reviewed by several independent researchers chosen by the journal) before it is revised and then accepted for publication. There are examples where scientists have gone straight to the media and it has later turned out that the research was flawed.* Furthermore, fenoterol was used around the world, and it was important that the paper was published in a scientific journal that was available around the world. No authority in Europe, for example, was going make decisions about the availability of fenoterol just because of a report published in a daily newspaper in New Zealand. So the paper had to be published in a good scientific journal, like the *Lancet*, and in the meantime we had to sit and wait, even though the deaths were continuing.

It was also clear that the Boehringer lobbying was having an impact on the New Zealand Department of Health. It was particularly galling for me to see the department's staff begin to sink under the wave of hostile reviews, since I knew that most of them had had no training in epidemiology. The few that had had any training had received it on my course at the Wellington Medical School – at that time I taught the only postgraduate epidemiology course in New Zealand. It was frustrating to see how little they had learned, and how easily they could be swayed by company pressure and trivial criticisms.

The one positive thing was that the paper had been accepted for publication in the *Lancet*. When it was published, we thought, at least the debate would be out in the open, and patients with asthma (and their doctors) would be able to make up their own minds about whether to use the drug.

The first indication that the paper might not be published after all came when Richard Beasley was discussing the study on the telephone with Malcolm Sears. As Beasley later recounted:

> He made some comment about 'if' the paper got published, and I told him that it had been accepted for publication. He said that there was a difference between having a paper accepted

---

* One such example is the cold fusion story, in which University of Utah chemist Stanley Pons and a British collaborator, Martin Fleischman, announced that they had created a sustained nuclear fusion reaction at room temperature. The discovery was considered so important that they announced it directly in the press, to ensure that they would get the credit for making the discovery, but their research has since been discredited. (G. Taubes, *Bad science: the short life and weird times of cold fusion.* New York: Random House, 1993).

and actually having it published. I wasn't actually aware of the distinction at that stage, so I asked him what he meant. He reiterated that there was a distinction, and just because the paper had been accepted did not necessarily mean that it was going to be published, but he didn't want to comment further.[35]

A few days later, on 14 March, I found Crane hurrying down the corridor with a letter from the *Lancet*, dated 3 March 1989:

In my letter to you of February 20, accepting the fenoterol paper, I mentioned the existence of critical reports commissioned by Boehringer Ingelheim. Some of those commentaries have now been sent to us and they are causing some anxiety. You will by now I hope have seen the long summary Fenoterol and Asthma Mortality by J D Wilson, so let me use that as a baseline.[36]

The letter then listed eight sets of questions about the study including the Asthma Task Force's claim that some of the control hospital admissions had not been for asthma, that some of the cases had been 'identified by the Task Force (the group providing the information) as not using or probably not using fenoterol before death', and that use of fenoterol 'might be a marker for severity rather than a cause of death in severe asthma . . . and your critics are not convinced by your attempts to allow for this confounder'. The letter concluded:

Three possible courses of action face a journal having second thoughts about a paper accepted but not yet published.
(A) Some argue that this case control study should be repeated with more carefully chosen and validated cases and controls and with information collected identically for both groups. Are you willing to withdraw the paper and do that work?
(B) Failing (A) you would be expected to counter convincingly the very critical comments of the Task Force (which, as they stand, invalidate the study) or to repeat the calculations after removal of the cases and controls upon which doubt has been thrown. Even then a franker discussion of the faults

in the study's design and execution would be necessary, as would a much more tentative conclusion.

(C) We will publish the paper, with such modifications as you feel able and willing to make, in the second half of the journal accompanied by a highly critical editorial in the same issue.

The *Lancet* does not normally shy away from controversy, but the foundations of this one are starting to look very shaky. We are only sorry to have had to raise such daunting obstacles so late in the progress towards publication.

The *Lancet* had buckled under the pressure, and had reneged on its acceptance of the paper. This was the top medical journal in the world, and it was not supposed to behave like this.

Crane called the *Lancet* that night, and was told that the journal's preferred option was for us to withdraw the paper, since they felt that we could not answer the criticisms. They commented: 'We've had many controversies, but we've never had anything like this'. Several lengthy faxes were arriving every day, and the journal had already received so much material that it had asked Boehringer not to send any more. What had particularly concerned the *Lancet* was the critical comments of the Asthma Task Force, which had been the main factor in its decision to rescind the acceptance of the paper. The journal wanted to put the paper 'on the backburner', and was in no hurry to publish it.

THE DEPARTMENT OF HEALTH DOES NOTHING

We were never able to get the Department of Health to hold Boehringer to account for what we regarded as a breach of the conditions under which the company had been given the manuscript. We met with George Salmond and asked him to write to the company stating that the department was concerned that the company had not kept the assurances it had given, and that the department would therefore not be willing to accept the company's assurances in future. Salmond said that he would send the letter, but it seems that he never did. In a subsequent letter to the Public Health Association, Warren Thompson, the department's Manager for Medicines and Benefits, stated:

It was alleged by the researchers that Boehringer Ingelheim had attempted to suppress publication of the research and the department did undertake to take the matter up with Boehringer. Both the Director-General and I had conversations (separately) with Dr J D Wilson, Medical Director of Boehringer. On each occasion, Dr Wilson was adamant that there had been no attempt by the company to suppress publication but that they had presented their point of view to the *Lancet*. That view was not supportive of the research findings.[37]

In fact, we had never alleged that Boehringer had tried to suppress publication directly; rather, we argued that the company had been given a copy of the manuscript on the condition that it would do nothing to upset the normal process of publication, and that it had violated that agreement by flooding the *Lancet* with hostile reviews. In effect, Boehringer sidestepped the issue of interference with publication, and instead denied any attempt to suppress publication (which we had never alleged). The department, which had previously assured us of its support, accepted the Boehringer denial.

A CRITICAL MASS

The next move by Boehringer was to get some of its reviewers together in a consensus panel. On 22 March 1989, we received a fax from the company inviting two of us to attend a workshop on our study on 1–2 April at the Ritz-Carlton, one of the most luxurious hotels in New York City. We had not previously been told about the meeting, and were being invited at very short notice. We replied:

> Boehringer Ingelheim was provided with a copy of the draft manuscript for limited purposes and on the condition that the company would not seek to influence the normal process of publication. The communications that Boehringer Ingelheim has subsequently had with the *Lancet* are a clear breach of these conditions. We consider that the proposed New York meeting would also breach those conditions and has the potential to be used by

Boehringer Ingelheim to interfere with the normal process of independent scientific review. We accept the importance of scientific discussion of our work and expect to participate in such discussions after publication. In general, we would expect such meetings to be independent of Boehringer Ingelheim.[38]

We had been invited to the meeting at the insistence of Anne Tattersfield, of the University of Nottingham. She was editor of the respiratory medicine journal *Thorax*, published by the British Medical Association, which had published our first laboratory study.[39] She had been invited to the meeting and was surprised to find that we had not. When an invitation was sent to us at the last minute, she still decided not to attend. Another colleague in the United Kingdom initially accepted an invitation, but withdrew after he found that he was to be booked on Concorde. Others accepted their Concorde tickets, went to the meeting, and by all accounts had a very nice time.

The meeting was chaired by Professor Walter Spitzer of McGill University. Other panel members were Dr Sonia Buist (Oregon Health Sciences University), Dr Peter Burney (University of London), Dr Ralph Horwitz (Yale School of Medicine), Dr Stephen Lanes (Epidemiology Resources Inc.), Professor Anthony Rebuck (Asthma Centre at Toronto's Western Hospital) and Professor Ann Woolcock (Royal Prince Alfred Hospital, Sydney). The meeting included those Boehringer reviewers who had made the most critical comments on our study, while some other reviewers who had made less critical comments were not invited or did not attend.[40] The Boehringer Ingelheim consensus panel concluded that 'The study design is seriously flawed and may lead to unjustified policy formulation and prescribing decisions'.[41]

The report from the Boehringer Ingelheim consensus panel listed dozens of criticisms of the study, almost all of which were incorrect, trivial or involved biases which would tend to produce false negative rather than false positive findings. The more serious criticisms fell into three groups:

(1) the study was based on prescribed medication, rather than the medication actually used;

(2) the findings could be due to 'confounding by severity' if fenoterol had been marketed specifically for people with more severe asthma; and

(3) we had obtained the drug information from different sources for the cases and controls, so the data might not be comparable.

All of these issues had been discussed at length in the paper itself. Good information on drug use was not available, although some Asthma Task Force members claimed otherwise. In any case, other studies had shown that almost all asthmatics use their beta agonist a great deal in a severe attack.[42] Although some of those who die (and some of those who do not) may not use their inhaler, any such problems with the drug information will tend to blur the effect rather than exaggerate it.

This is one of the fundamental concepts of epidemiology,[43] but it is not well understood by researchers in other fields. Let us suppose that there is a drug (fenoterol) that doubles the risk of death, compared with the standard treatment (salbutamol). If we do a study of 1000 people taking salbutamol and 10 of them die, and we also study 1000 people taking fenoterol and 20 of them die, then the 'relative risk' is 2.0. However, if there is random misclassification of exposure (i.e. for some of the people who are actually taking salbutamol we mistakenly think that they are taking fenoterol, and for some of the people who are actually taking fenoterol we mistakenly think that they are taking salbutamol), this will 'mix up' the two groups and the relative risk that we see will be lower than 2.0 (but still higher than 1.0). Such misclassification will almost always tend to blur the differences between the two groups and produce 'false negative' findings; it will rarely produce 'false positive' findings.

The claim that fenoterol had been marketed for use by more severe asthmatics was made many times, and continues to be made today. It is also argued that fenoterol was a 'new' drug, and that it might therefore have been used by more severe asthmatics who had been obtaining insufficient relief from other drugs. In fact, no evidence has ever been produced that fenoterol was marketed for more severe asthmatics, and there is certainly nothing in the advertisements at the time, or in anecdotal reports, to support this.[44] Nevertheless, the 'fact' that our findings were due to confounding by severity became widely quoted in the following months.

The third criticism, that we had obtained the drug information from different sources for the cases and controls, was more valid and was the only major flaw in the study. We had made the same criticism ourselves in the paper.[45] There was no other way of doing the study. However, we did carry out several validation exercises, along with various analyses designed to find out whether serious bias was occurring because of this problem. All of these showed that it was very unlikely that the study findings could have been caused by such a bias. For example, the easiest way to check this is to restrict the analysis to cases and controls who were reported to have exactly one type of beta agonist inhaler – that is, to exclude those who were reported to have had more than one, and those who had none. When I did this, the findings did not change. We also did a validation study of 48 patients discharged from Wellington Hospital in 1988 with a diagnosis of asthma. We recorded their prescribed medication from the hospital notes, and also asked their general practitioners what their patients' prescribed medication had been (using the same procedure as in the Asthma Task Force survey). There was very good agreement and no evidence of a systematic bias.[46]

INDEPENDENT REVIEWS

By now it was becoming clear that to some extent we would have to play the company at its own game. We did not have the funds to commission reviews, and we would not have done so in any case. However, the manuscript had been informally reviewed for us by several colleagues, including Robert Beaglehole, Robert Scragg and Sir Richard Doll, as well as the members of the Department of Health review panel, and some colleagues at the IARC in Lyon (John Kaldor, Peter Boyle and Rodolfo Saracci).

These reviews all raised some criticisms, but were generally positive. Doll's remarks were typical. He commented on the problem of comparability of information between cases and controls, but essentially supported the study's findings:

As I am sure you are aware there are worries about the design of your case-control study . . . Having said that, I have no doubt that you have made a strong case for the hypothesis that fenoterol is responsible for many of the deaths, but I can't say anything further without more detailed knowledge of the epidemic.[47]

We sent these reviews to the *Lancet*, with the permission of the authors, and asked other colleagues to write to the *Lancet* independently to support the publication of the study. For example, a further review was sent independently by Professor Allan Smith (University of California, Berkeley), my old PhD supervisor, who argued:

One can always list flaws in studies such as these. The question is whether they invalidate the findings to the extent that they should be suppressed. History cautions against rejecting epidemiological studies because of perceived flaws or biases. I will give but two examples, both of which I am sure you are well aware of. The initial studies of smoking and lung cancer were widely criticized as flawed and biased. Indeed, tobacco companies were able to generate criticisms for decades. The first finding of a relationship between thalidomide and phocomelia was perceived as flawed and the report of it from Australia was rejected by the *Lancet*. We are not faced here with a thalidomide disaster. However studies such as this one should see the light of day. Boehringer would be better advised to support studies to investigate the issue further, rather than attacking a study because of its possible biases.[48]

THE *LANCET* GIVES WAY

In a 12-page letter faxed to the *Lancet* on 17 March 1989, we discussed the Boehringer and Asthma Task Force criticisms in detail, and stated:

As you will have anticipated, we were disturbed by your letter of 3 March 1989. We are concerned that our scientific paper, which was accepted unconditionally . . . could have this unconditional acceptance withdrawn following submissions made by a pharmaceutical company . . .

We recognize that the company has placed the *Lancet* in a difficult position. However, we can counter convincingly the very critical comments of the Task Force. . .

We would like to commence by outlining some of the steps which our case-control study had been through prior to being submitted for publication in the *Lancet*. It should be noted at the outset that, although we have not set out to commission reviews in the manner that Boehringer has, our study has received many favourable reviews from independent epidemiologists (copies enclosed) . . .

The recent negative comments from the Asthma Task Force are disturbing in light of the Health Department Review Panel meeting in December, and the Task Force's acceptance (at that time) of its findings. We have spent a great deal of time with members of the Task Force (both privately and under the auspices of the Health Department), going over their criticisms, accommodating the important ones, and explaining why the rest were incorrect or trivial. We had thought that most of these difficulties had been resolved by the Department of Health Review Panel, particularly when the Chairman of the Task Force accepted the Review Panel's endorsement of our study design. It is thus distressing to see these same criticisms raised again by the Task Force, and also appearing in the Boehringer submissions. Fortunately, we can answer them, as we have done many times before . . . .

We realize that the actions of the company have placed the *Lancet* in a difficult position. However, we welcome the opportunity to clarify these issues by expanding our paper. We look forward to seeing it in print.[49]

We called the *Lancet* on 30 March, and learned that it had accepted these arguments, and would publish the paper after all. We later sent a letter for publication in the *Lancet* stating:

We would like to express our concern at the attempts by Boehringer Ingelheim to influence the normal process of publication of a scientific study, and to influence its subsequent assessment. We consider that such actions have major implications for the relationship between pharmaceutical companies and the medical profession . . .

Prior to submission to the *Lancet*, the study was extensively reviewed by internationally respected expidemiologists who concluded that the study was well-designed and that the findings justified public health action. It was for this latter reason that we took the step of providing the New Zealand Department of Health with a copy of the manuscript at the same time as we submitted the manuscript to the *Lancet*. To enable Boehringer Ingelheim (the company that manufactures fenoterol) the opportunity to consider the withdrawal of fenoterol from New Zealand, we allowed the Department of Health to provide Boehringer Ingelheim with a copy of the manuscript, but on the written condition that Boehringer Ingelheim did not attempt to interfere with the normal process of publication. Boehringer Ingelheim subsequently submitted direct to the *Lancet* a considerable amount of material that made a selective and highly critical attack on our study; thereby contravening the conditions by which they accepted the manuscript. This, on referral to us was rebutted and publication proceeded, but we are concerned at the precedent that Boehringer Ingelheim has set . . . It is possible that a journal without the independence and integrity of the *Lancet* might have been influenced by these attempts.[50]

Despite our attempts to let the *Lancet* off the hook, the journal resisted the notion that anything untoward had happened. In reply, Deputy Editor David Sharp claimed:

We were not put under pressure but merely asked if we would like to see copies of reports commissioned by the company. As you know our late doubts related more to what the Asthma Task Force had to say than anything else . . . I propose to put your letter on ice since I think you will be well advised to think again about it.[51]

Although the journal was reluctant to accept that anything unusual had happened, at least it was again prepared to publish the paper. Fortunately, Richard Beasley was briefly visiting London and went to the *Lancet* offices, where he met the editor, and saw 'the first box' of material that the journal had received from Boehringer. More importantly, he was able to review the proofs – the draft copy of the printed version of the paper – so that the typing and numbers could be checked. These are usually sent by airmail or by fax, but we had not received them. It later turned out that they had been sent by seamail. If Beasley had not been visiting London, and had not gone in person to check the proofs at the *Lancet*, we would have waited several months for them to arrive. During that time there would have been more criticism of the paper and it might never have been published. But now we had checked the proofs, and the publication date was set as 29 April 1989.

CHAPTER 7 **Adverse reactions**

BOEHRINGER LOBBYING CONTINUES

Before the paper was published, the Boehringer lobbying continued. The Department of Health seemed increasingly nervous
and cautious, and surprisingly friendly with the company. On 15
March 1989, Salmond wrote an internal departmental memorandum stating:

> Doug Wilson left me with the attached press releases he and
> Boehringer have prepared for release in the event of a media
> expose. Doug is keen to work in with us over press releases
> and other contact with the media and using the opportunity
> for some good asthma education. In that regard I think Doug's
> background is excellent.[1]

Although this joint approach did not proceed, the department
seemed very reluctant to make any concession that there could
be a problem with the drug. The then Minister of Health, Helen
Clark, later commented:

> [The department's] attitude to the research was clearly somewhat
> sceptical. A draft press release prepared by the department
> referred to the study associating the drug with users' deaths and
> then quoted the Principal Medical Officer for Medicines and
> Benefits, Dr Riseley, as saying: 'This is unproven. If there is a risk
> with the asthma medicine fenoterol, it is an extremely low risk'
> . . . This was an extraordinary statement to make at that time,
> and certainly appeared to prejudge the issue for the department.[2]

After criticism of this press release, the department modified it
and agreed to send out a copy of the abstract of the paper with a
letter to doctors. The draft letter stated:

The Department and the Medicines Adverse Reactions Committee have sought and received advice from a number of experts. This is conflicting and the situation may not be clarified for some time. The article does suggest that there may be an increased risk of death from asthma in patients prescribed fenoterol by aerosol who have severe disease, or who also use oral steroids. In these circumstances it would be prudent to consider whether the use of fenoterol in such patients should be modified. Practitioners are advised to take this into account when prescribing for these groups. [3]

Boehringer was sent a copy of this draft, and replied by fax to the department, stating:

We are most concerned at your intended communication with doctors . . . Enclosing a copy of the abstract suggests Departmental approval . . . unless the Department's approach is modified we are left with no option but to take every step available to us to protect ourselves. We must discuss the subject with you immediately and certainly before any communication from the Department goes to the medical profession.[4]

Boehringer's head office in Germany also contacted the department:

We have learned that the content and ambiguity in your letter has caused substantial anxiety and confusion among asthma patients and doctors . . . the responsibility for the medical and legal consequences rest with you . . . We expect [your next] letter to state that therapeutic conclusions cannot be drawn from the Crane et al study in order to resolve anxieties that have been created among asthmatic patients and their physicians in New Zealand.[5]

Salmond advised the department not to circulate the abstract of the *Lancet* paper, but this decision was reversed in the week after publication, when a complete copy of the paper was sent to all doctors in New Zealand.

The Department of Health's weak response to the company's lobbying was not helped by the fact that MARC did not include an epidemiologist. The committee was chaired by Gavin Kellaway, of the University of Auckland, and included Professor Ralph Edwards, Head of the Toxicology Group at the University of Otago, which monitored adverse drug reactions. During this period, MARC reports consistently tended to downplay the findings not only of the epidemiological studies, but also of the laboratory studies. For example, a MARC statement, issued just before the publication of the paper, claimed that fenoterol was chemically similar to salbutamol and other bronchodilator medicines used in asthma, and that its adverse reactions profile was similar to those of salbutamol and other bronchodilator medicines. This statement ignored not only the experimental evidence that fenoterol has greater cardiac side-effects than salbutamol, but also the adverse reaction reports from Edwards's own unit, which showed 'a higher reporting rate for death and cardiac adverse reactions with fenoterol as compared with salbutamol'.[6]

THE ELWOOD AND SKEGG REVIEW

Although MARC did not include an epidemiologist, the Department of Health had commissioned a lengthy review of the study by epidemiology professors Mark Elwood and David Skegg at the University of Otago. This, like the other independent reviews, was quite different in style and conclusions from the reviews commissioned by Boehringer Ingelheim. It contained some criticisms of the study, and particularly of the use of different information sources for the cases and controls, but it recommended

> that action is taken on the basis that this study suggests there may be an increased risk of death from asthma in patients prescribed fenoterol by metered dose inhaler who have severe disease . . . This evidence, although far from conclusive, suggests that the current practice in the use of fenoterol should be modified. Unless there are advantages to fenoterol which can

be documented, the evidence from this study (even though insecure) argues for the use of such alternatives rather than fenoterol, or for other changes in the way MDI fenoterol is used.[7]

These recommendations touched on a key public health issue. No one had produced any evidence that fenoterol was better than the other commonly used beta agonists. At best, it was similar to them; at worst, it was much more dangerous and could be causing an epidemic of deaths. Thus, the prudent course was to restrict the use of fenoterol until the issue had been sorted out.

However, MARC had already met and made its decision before the Elwood and Skegg review was sent to its members. The committee remained reluctant to take any action, or make any statement, which might cast doubts on the safety of the drug.

WAITING FOR PUBLICATION
In any case, the waiting for publication was nearly over. On the Saturday afternoon one week before the paper was published, I stopped at Richard Beasley's house for a beer on the way home from work, and we wondered what the coming weeks would bring. Although we were confident that the study findings were valid, the months of hostile reviews and criticism had taken their toll. Some good advice had been received in a fax from Stephen Leeder, chair of the Department of Health review panel:

> I have read [the critiques of the case-control study] and do not feel greatly moved by them, I must say. Virtually all of them raise concerns that I felt were ventilated in our meeting with you prior to Christmas, and dealt with as satisfactorily as one could . . . many of the concerns seem to take their origin from points other than in your paper . . . I realize that this is a very upsetting time for you and that you will be attacked almost as much on personal grounds as well as ones relating to the quality of the science . . . I'd suggest you take the family, your portable CD, some Scotch, and go for a sail around the Bay of Islands.[8]

Unfortunately, we couldn't take Leeder's advice, but at least the debate was now going to occur in the open rather than behind closed doors, and people with asthma, and their doctors, would be able to make up their own minds about the issue.

The next morning I was woken by a telephone call at 6 am. My mother had had a heart attack. She was 72, would be in hospital for several weeks, and would take months to recover. My father, who was 85 and was also seriously ill, would have to be cared for at home initially, and then moved to a rest home for several months. People kept commenting that I looked tired and stressed and that the fenoterol controversy must be very difficult for me. But the stress from my parents' illnesses was much worse than the stress at work. My father died nearly a year later, and during this time my family situation took up most of my free time. This limited my ability to respond to the attacks on the fenoterol studies.

PUBLICATION

The publication date had been set for Saturday, 29 April.[9] We preferred to make no press statements on the issue, and to restrict ourselves to replying to any criticisms in the scientific journals. However, we had previously agreed to give interviews on the day of publication to several reporters, who had previously threatened to break the story before the paper was published. We planned to give these interviews and then stay out of the media.

Richard Beasley had spoken to the *Lancet* twice, and each time he had been told that the publication time was midnight GMT on Friday, 28 April, which was midday on the Saturday in New Zealand. We advised the Department of Health of this, and it prepared to issue a press statement on the Saturday. The Minister of Health, Helen Clark, had been advised by the department that 'Boehringer Ingelheim have a vigorous media campaign prepared. It is not known when it will be launched'.[10]

On the Friday, 28 April, the story hit radio and television. Boehringer New Zealand held a press conference at noon and issued a publicity package to reporters. At the press conference, Doug Wilson said that the company would continue to market fenoterol: 'He urged patients to make no moves over the drug on the basis of a difference of opinion between physicians . . . with regards to interpretation of this study. He said it would be surprising if some doctors moved away from the use of [fenoterol].'[11]

Colleagues also began to call us to let us know that they had been personally couriered a bulky publicity package. This package, the first of many over the next two years, was couriered to almost all doctors, pharmacists and health reporters in New Zealand, and was also delivered to many respiratory physicians in other countries. They received the package before they had received anything from the Department of Health.

The package highlighted the conclusions of the New York meeting of Boehringer reviewers that the study had 'serious flaws in design, execution and analysis which rendered its results uninterpretable'.[12] It also contained many pages of the same misleading graphs that had been submitted to MARC, along with tables and other information which it claimed demonstrated a lack of association between fenoterol sales and asthma deaths. A follow-up press statement on the Saturday stressed that no other health authorities had taken action on fenoterol, although, of course, none of them had had time to respond. It stated 'Berotec is only one drug the Company manufactures and if it had any doubts about Berotec's safety profile, would not hesitate to withdraw it immediately'.[13]

The most effective Boehringer weapons were the reviews it had commissioned, and the report of the 'consensus meeting' in New York. A report in the Wellington morning newspaper, the *Dominion*, stated: 'The New Zealand study on asthma deaths had "serious flaws in design, execution and analysis which rendered its results uninterpretable", an international advisory panel convened by the drug's manufacturers has said.'[14]

This report was unusual in that it listed all the members of the panel, and mentioned that it had been convened by Boehringer. Most newspaper reports did not, so members of the public did not realise that the consultants had been hired by Boehringer. Also, people were impressed by headlines such as 'Asthma study flawed, say foreign experts'. For example, a report in the *Auckland Star* stated that 'Experts from the United States, Canada and

Britain have attacked a New Zealand study which linked the asthma drug fenoterol with an increased risk of death, saying that it was "seriously flawed" and could lead to unjustified prescribing decisions'.[15] After all, these experts had met in New York, and we were only from Wellington.

Boehringer later denied having pre-empted the publication of the paper, and they were supported in this by the *Lancet* which stated that the embargo had been lifted at midnight GMT on Thursday, 27 April – that is, noon on Friday, 28 April New Zealand time. This was one issue where Boehringer got it right and we got it wrong. The *Lancet* was right to tell Beasley that the publication date was midnight GMT on Saturday, 29 April. However, it did not tell him – or maybe it did and he didn't understand – that the embargo on reporting the study was one day earlier. Because of this misunderstanding, or misinformation, Boehringer was able to pre-empt both our group and the Department of Health, and mount a massive publicity campaign on the Friday. As a result, many doctors and members of the public had already made up their minds by the time they heard our response the next day.

Inevitably, we were drawn into responding to the attacks from the company, and there was a great deal of media coverage over the following weeks. The company then attacked 'media hype' for giving bad publicity to the drug. The Managing Director of Boehringer Ingelheim New Zealand, Ron Scobie, stated:

> We have heard of a case over the weekend where a patient had been taken off the drug and ended up in Auckland Hospital after an attack. If there was a real risk with the drug we would have taken the necessary steps to withdraw it. My eight-year-old daughter takes [fenoterol] and I have no plans to take her off it.[16]

Many more Boehringer publicity packages were to follow, and were once again delivered to virtually all doctors in New Zealand. Many Boehringer reviewers also issued individual press statements. For example, the following statement appeared in the *Australian Doctor*:

Professor Ann Woolcock, Professor of Respiratory Medicine at the University of Sydney, said that there had been no association between the drug and asthma deaths in Australia. Professor Woolcock warned that patients should not stop taking any asthma medications in reaction to this report.[17]

Similar comments were made by Charles Mitchell (University of Queensland) and Paul Seale (University of Sydney). In fact, it was impossible to know whether or not there was any association between the drug and asthma deaths in Australia, because the market share was too low (less than 5 per cent) for any increase in deaths to show up in the time trends, and no one had done a case-control study as we had in New Zealand.

The Boehringer publicity campaign was very effective, but at least some doctors were sceptical. For example, one doctor from Dunedin wrote to Boehringer on 13 May:

> I found your information unhelpful, for several reasons . . . The data you provide do not, as you suggest, refute the hypothesis that unsupervised administration of fenoterol by inhalation in severe asthma increases the risk of death. Your data seems weaker than that of Crane et al, yet you use it to draw a dogmatic conclusion . . .
>
> Considerable sums have been invested in convening a meeting of an international advisory group for two days and sending every registered medical practitioner in New Zealand two mailings of information. Yet there is no indication of any intention to support a serious attempt to replicate the findings of Crane et al, nor to support them to extend their research and rectify any shortcomings. This would seem to be a more responsible way to deal with results that you do not wish to believe.[18]

ASTHMA TASK FORCE

The Asthma Task Force also issued a critical press statement on the Friday before the paper was published. This was followed up with statements which were sent out as part of a special Boehringer publicity package on 1 May. In a later newspaper interview, Malcolm Sears was asked why the task force was criticising the study when the study design had been approved by the Department of Health review panel, which concluded that action should be taken on the basis of the findings. Sears disputed this:

> The meeting tried to find common ground and to identify any areas that still needed to be explored. We were left with considerable disagreement. Professor Beaglehole has been quoted as saying the issues were explored and resolved. But they were not resolved. Substantive issues remained on which we had concerns.[19]

The 'substantive issues' raised by Sears related to the study design and the selection of controls – issues that had been resolved at the review panel meeting.

Much task force material was used by Boehringer Ingelheim under other guises. For example, one of the first Boehringer publicity packages quoted some analyses that Harry Rea had done of patient records at Green Lane Hospital. The publicity package stated that the data had been obtained from the Department of Health, when in fact they were taken from a task force submission to the department. The wording in the Boehringer publicity package was so ambiguous that these data were subsequently reported as being from 'a Health Department study'.

WELLINGTON MEDICAL SCHOOL

The Asthma Task Force's support for the Boehringer publicity campaign was also causing a stir at the Wellington Medical School. As chair of the task force, Tom O'Donnell had issued a press release criticising the research, which, of course, had been conducted in his own medical school. The Public Health Association arranged for us to give a lunchtime seminar on the Tuesday after the paper was published. The large lecture theatre was packed with staff and students. A meeting of heads of departments was abandoned because of the lack of a quorum. We presented our studies and didn't comment on the politics, but at the end Don Matheson (now Deputy Director-General at the Ministry of Health) was applauded when he stood up from the audience and criticised the Boehringer publicity campaign and the dean's participation in it.

Similar comments were made several weeks later in a more subtle manner by the chair of the Medical School Council, William Scollay, in his annual address. In front of the Minister of Health, together with O'Donnell and the other professors in formal academic dress, he stated that the fenoterol study was a good example of the type of research that the school should be doing:

> Now the authors of this study may be wholly right, only half-right or otherwise, but the point is the questions have been asked, and have become the subject of informed and public, even international, debate. The partnership of the teaching hospital and the Medical School ensures that we have an environment in which questions will be asked, to the benefit of all New Zealanders.[20]

In a subsequent interview in the New Zealand *Sunday Times*, O'Donnell referred to some of the criticism:

> I think that there is a risk, as there has been, in my being seen to be a defender of a drug rather than a clinician analysing a particular study . . . [The study] has raised another possibility which should be pursued . . . [However], I'd have to say that we haven't discussed [any plans to pursue it].[21]

The situation at the medical school was later commented on by Warren Thompson, the Department of Health's Manager for Medicines and Benefits, who was quoted by the *Christchurch Star* as noting that

> There has been a very definite element of injury as [members of the task force] see it to their reputations and it has reached quite difficult proportions, particularly within the Wellington Medical School. The Asthma Task Force may feel under attack . . . but

they also feel that they are right, that the young turk researchers have not made their case properly.[22]

DEPARTMENT OF HEALTH

In comparison with the Boehringer publicity, the Department of Health's effort was feeble. By the Monday after the publication of the paper, most doctors had received a one-page letter by regular mail advising that 'Fenoterol will not be withdrawn from the market, but doctors should review and perhaps modify the treatment of severe asthmatics'.[23]

The Minister of Health, Helen Clark, later commented:

> For reasons not clear to me at the time, but in retrospect because of sensitivity to Boehringer Ingelheim's opinion, the Department had decided not to send the abstract of the researchers' paper out to doctors. Thus all that was received by the latter was a one-page letter urging caution in prescribing. Belatedly in early May the Department decided to send out copies of the paper, but undoubtedly it could have assisted medical practitioners even more by also sending at least a summary of the Elwood/Skegg review. Its actions on the issue, which after all involved public safety, could at best be described as minimal.[24]

THE PUBLIC HEALTH RESPONSE

Public health researchers were less nervous about rebutting the Boehringer publicity. Elwood and Skegg issued a press release:

> Since the publication of the study on fenoterol and asthma deaths last Saturday in the *Lancet*, every doctor in New Zealand has received extensive material on the topic from the manufacturers of the drug, Boehringer Ingelheim, and in contrast has received a short and imprecise letter from the Department of Health, commenting that the evidence they have received is conflicting and that they want to seek further international opinions. We are concerned that this imbalance of information reaching ordinary doctors makes it very difficult for them to make appropriate decisions.

In early February, we were asked by the Department of Health to undertake a detailed and objective review of the paper which has recently been published, and in addition to review over 400 pages of comments relating to that study provided to the Department of Health by many experts both in New Zealand and overseas. We were asked to do this because of our independence from both the group performing the study and the manufacturers of the drug . . . While of course we do not claim that our review covers every aspect of this extremely difficult question, we consider that the Department of Health has the information required to give much more definitive guidance to doctors.

The recommendations in our report included the following:

We recommend that action is taken on the basis that this study suggests that there may be an increased risk of death from asthma in patients prescribed fenoterol by metered dose inhaler who have severe disease, or who also use oral steroids. This evidence, although far from conclusive, suggests that current practice in the use of fenoterol in such patients should be modified. Individual physicians and professional committees must balance this conclusion with any cogent arguments for the use of fenoterol rather than alternatives in these patients.[25]

Elwood and Skegg submitted a similar letter for publication in the *Lancet*, but this was rejected. The *Lancet* did publish critical letters from the task force and from the Boehringer consensus panel, as well as our reply.[26]

Robert Beaglehole was particularly blunt in his condemnation of the Department of Health:

The Professor of Community Health at the Auckland School of Medicine, Robert Beaglehole, has accused the Health Department of possibly risking lives while it waits for the 'big boys' of international medicine to report on a controversial asthma drug. Dr Beaglehole said it was unlikely better research data would be provided on the drug fenoterol than was already available . . .

Dr Beaglehole said he could not understand, what he terms, the department's 'neither confirm nor deny' attitude. 'There is no point in procrastinating', he said. 'We have to act on the available information.'[27]

Rod Jackson made similar comments:

When you are dealing with a life-threatening disease and serious questions have been raised about the safety of a treatment, then it must be considered guilty until proven innocent. The reaction of Boehringer and the Task Force would suggest that the drug is innocent until proven guilty. I do not believe that is acceptable when there are alternative medications available.[28]

Similarly, a report from the Public Health Association stated:

Sales of fenoterol should not take precedence over science . . . Boehringer Ingelheim appeared to pre-empt constructive assessment of the research by circulating its own promotional material before doctors had read the original report in the *Lancet* . . . The research findings are disturbing to asthma sufferers and they need to be checked as quickly as possible by researchers independent of those with vested interests . . . Clearly the pharmaceutical company and those who appear to have aligned themselves closely with its views would not be the most appropriate people to contribute further to research on the safety of asthma medication.[29]

The Public Health Association was particularly concerned that Boehringer had continued to advertise and promote the drug after our findings were made available to the company, and even after the paper was published. It called for an inquiry to review:
(i)   the interrelationship between the company and the Department of Health;
(ii) how the Department of Health and Boehringer-Ingelheim responded to the findings of the study;

(iii) the activities of Boehringer Ingelheim to promote the use
of fenoterol from the time the safety of the drug was first
questioned; and

(iv) the promotional activities undertaken by all pharmaceutical companies in relation to medical research and
medical education.[30]

The call for an inquiry into the actions of the Department of
Health was particularly courageous since the Public Health Association had only recently been established and was funded largely
by Department of Health contracts. The association also stated:

It is time that clear limits are set on promotional practices
by pharmaceutical companies and that these are monitored
by the Department of Health. Appropriate action should be
taken when the limits are breached. . . Limits are essential to
ensure that there is no risk that the promotion and sales of
drugs could take precedence over patients' health or that the
nation's pharmaceutical bill could bear the cost of marketing
excesses.

An editorial in the *Sunday Star* commented:

New Zealand is witnessing an astonishing dispute in the medical
community over the use of the drug fenoterol . . . The Asthma
Task Force, comprising four senior clinicians, have attacked
the study's methodology and findings. Not unsurprisingly,
Boehringer Ingelheim, the drug company that manufactures
fenoterol, has joined a spirited campaign against the study.
Meanwhile, asthma sufferers are left totally confused . . .
    Even to a layman's eye, there appears to be sufficient expert
opinion raising serious questions about the use of the drug.
Where such doubts exist at all, it would surely be prudent to
err on the side of caution . . . The department should advise
suspending fenoterol use in the at-risk categories until the
argument is settled decisively.[31]

In some parts of the country, many doctors were impressed by the Boehringer publicity, and sales of fenoterol had been falling only very slowly. It was clear that fenoterol was no better than other available drugs, and might be much worse, but doctors were still keeping their patients on it. We heard many reports of patients who had asked to be taken off the drug, but had been persuaded to stay on it by their doctors. Many doctors were expressing sympathy for the company rather than for their patients, and were concerned that the New Zealand branch might have to close down if the drug were withdrawn. The most common comment was along the lines of 'What a pity that this has happened to such a nice company which has done so much to support research in New Zealand'. I even heard people saying, 'You don't bite the hand that feeds you.' The only criticism of drug companies that I heard from doctors was of a rival company, Edinburgh Pharmaceuticals, which had placed advertisements to remind patients that its product, salbutamol, did not contain fenoterol.[32] It was very rarely that I heard a respiratory physician or general practitioner express concern for the patients who were using the drug.

It was hard to resist the conclusion that most doctors, and Department of Health officials, did not really understand the issues and were making decisions on the basis of what 'influential people' thought, or what was best for their own careers. The company's many years of advertising, funding of research and conferences, and regular contact with doctors were paying off. The doctors had for years relied on the drug companies as their main source of prescribing information, whereas the Department of Health was seen as weak and ineffectual.

Some doctors may also have been influenced by rumours. We had already learned of false rumours that the study had been done without a protocol, that the protocol had been changed, or that some of the controls did not have asthma. As the controversy grew we began to hear more unpleasant rumours: that I was motivated by politics; that the study had been done without ethical approval; that Sir Richard Doll had been acknowledged in the *Lancet* paper without his consent; that the data from the Auckland case-control study had been used without permission; that the study had been funded by a rival drug company; and so on.

These rumours never appeared in print, so it was difficult to refute them. At one stage I was called by a newspaper reporter who had been told 'confidentially' by a member of MARC that the study had been funded by another drug company. I was able to set the record straight, but I wondered how many other people had been told the same thing but had not called me to check. The rumours clearly struck a psychological chord in many people who were only too keen to believe them. When unwelcome news comes, it is natural to want to 'shoot the messenger' and to go through a period of denial. It seemed that denial was psychologically easier if you could tell yourself that the researchers had dishonest motives.

FEAR OF FLYING

The debate with the company and the task force continued for many months, and involved letters to medical journals, statements in the newspapers and television news reports, as well as the frequent Boehringer publicity packages. One interesting story which emerged concerned the studies of fenoterol and baboons in South Africa. The results had been published in 1972, before the drug was marketed internationally.[33] Baboons were chosen because they are very similar to humans with respect to drug side-effects. Three of the eleven baboons died after receiving doses of fenoterol similar to those that had been used in humans.[34] Doug Wilson claimed that Boehringer had never heard of the study,[35] which was peculiar, since the company had funded it. Subsequently he claimed that the senior author of the study had concluded that the deaths of the baboons were due to the anaesthetic used, rather than to fenoterol. However, this explanation had not been mentioned in the report of the study, and the deaths were not characteristic of anaesthetic deaths. If the deaths in the first study had been due to the anaesthetic used, then the follow-up study should have involved a comparison of the effect of fenoterol and the anaesthetic, alone and in combination. Instead, a second study was done in 30 baboons, with a different anaesthetic and fenoterol given by a different route.[36] Two more animals died after being given fenoterol, but the study was still subsequently claimed by the company to 'exonerate' fenoterol. It also came to light that fenoterol was also sometimes given to pregnant women to delay labour, but that it was 'recognised as having toxic effects . . . [and] could cause abnormal cardiac rhythm and increase the heart rate'.[37]

We did not want to enter into a campaign against the company. However, no one else knew all of the evidence, and therefore we felt obliged to respond to some of the more outrageous claims made by the company, which may have seemed plausible to readers who were not fully aware of the evidence. Once again, no one was willing to let the other side have the last word, and this led to an increasingly vitriolic correspondence between us and the Task Force in the *New Zealand Medical Journal*.

I must confess to being particularly pleased with one letter that I wrote to the *New Zealand Medical Journal*. This was a reply to yet another letter which claimed that the association of fenoterol with asthma deaths was just an 'artefact' akin to finding a spurious association between insulin use and death from diabetes. We responded to this in a letter entitled 'Fenoterol and fear of flying':

[It] is quite correct that a study of death in diabetes might find a spurious association between insulin use and death from diabetes. Similarly it would not be difficult to demonstrate (in a badly designed study) associations between prescription of certain classes of asthma drugs and asthma deaths, or air travel and death in a plane crash. However, [it is incorrect to believe that this] . . . has any relevance to our fenoterol findings. Our studies have examined chronically severe asthmatics, virtually all of whom have been prescribed inhaled beta agonists. We have found that patients prescribed one drug within this class (fenoterol) have a much higher death rate than patients prescribed other drugs within this class. This is analogous to finding that travellers on one airline have a much higher death rate than travellers on other airlines. The appropriate conclusion is not that we are observing a spurious association between air travel and death in a plane crash, but rather that some airlines are safer than others. Prudent air travellers who study the safety statistics when given the choice would prefer to fly Qantas and would avoid certain other airlines. Prudent asthmatics and their doctors would be advised . . . to use alternatives to fenoterol.[38]

At first, I had wanted to compare taking fenoterol to flying on Aeroflot, but we were concerned that the airline might have considered this to be defamatory. This letter was typical in that I wrote the draft, but all four members of the group contributed to it, and in the final version someone else was the first author. Academics are usually fiercely competitive about authorship and will compete fiercely to be first author of a publication. The fact that we essentially took turns to be first author of our various publications reflected the camaraderie, and the 'siege mentality', of the group during this period, which meant that the usual rules did not apply.

MEDICAL RESEARCH COUNCIL

One official body that was less than helpful during this time was the Medical Research Council.* Most of the leaders in New Zealand medical research had been to Otago Medical School together. This was the only medical school in New Zealand until the Auckland school was established in the 1960s; it had also established clinical schools in Christchurch and Wellington. The deans of the four medical or clinical school were all members of the MRC Council. Furthermore, the Asthma Task Force had been formed and funded by the MRC, and it was clear that some

---

* In 1990, the year after the fenoterol story hit the headlines, the Labour Government, with Helen Clark as Minister of Health, passed legislation to change the Medical Research Council into the Health Research Council (HRC). The change saw an increased emphasis on epidemiology and other public health research, rather than traditional biomedical research, and people with a wider range of backgrounds and disciplines were appointed to the main council and the various committees. I was on the Public Health and Māori Health committees from the inception of the HRC in 1991 until 1996, and was on the main council in 1993–96.

council members felt that criticism of the task force was criticism of the MRC itself. Thus, although a few members of the council were very helpful, we generally felt that we were coming up against 'the establishment' and not getting a fair hearing.

Of particular concern was the MRC's support for the task force's demand for an inquiry into our study. This had originated from a call by the New Zealand Medical Association for an inquiry into the safety of fenoterol, and a similar call by the Public Health Association for an inquiry into the way that the Department of Health had handled the issue. The Task Force then issued its own press statement demanding an inquiry into our study, rather than into the safety of the drug or the conduct of Department of Health. Malcolm Sears was quoted in the *Auckland Star* as saying: 'It is very important to have an independent expert review. It will take weeks rather than days, maybe even months . . . [It] is a matter for the Medical Research Council. They must decide how the review will be done.'[39]

The MRC decided to hold its own preliminary meeting on 25 May to discuss the study. In our submission to the meeting we argued:

> Following publication of our paper, the Department of Health
> has issued careful and responsible recommendations which
> closely follow those of Elwood and Skegg. We are concerned
> that Boehringer Ingelheim and the Asthma Task Force have
> issued dissenting views and thus caused unnecessary confusion
> for asthmatics.
>
> We are also concerned by the public call by the Asthma Task
> Force for a further investigation into our study . . . our work
> has been extensively reviewed by independent groups on two
> occasions, as well as being extensively reviewed by the *Lancet*.
> Submission of the manuscript was delayed for six weeks to enable
> a joint review of the basic data by our group and members of the
> Asthma Task Force. Both reviews were commissioned by the
> Department of Health, but we were assured that these reviews
> also had the approval of the MRC. The second major review
> (by Elwood and Skegg) concluded that 'we are not convinced

that further reviews of the data collected for this study will be helpful'. We thus regard the action of the Asthma Task Force in calling for a third review as demonstrating bad faith and an unwillingness to accept the findings of the extensive reviews which have occurred already.[40]

The meeting was chaired by the deputy chair of the MRC, who was a lawyer, and included the various council members (including O'Donnell), as well as Mark Elwood, Ralph Edwards, Harry Rea and Malcolm Sears, along with Richard Beasley and me. I felt that the only way to get our point of view heard was to state it fairly aggressively. Although some council members reacted negatively to this, it did seem to make them take notice that we were unhappy about what was happening and were not prepared to have a 'kangaroo court' inquiry into the study. One factor which may have influenced the council's eventual decision not to proceed with a further review of our study was that I mentioned, for the first time, that we had done a second study. This had used a different design which avoided the problems with the first study, but had produced similar findings. The council thus accepted the futility of holding a third inquiry into the first study. Instead, it issued a vague press statement which essentially said that the evidence was inconclusive.[41]

Thus, the first round of the fenoterol controversy came to an end with a win on points to Boehringer Ingelheim. But the second round was about to begin.

It is very hard to know how to respond, as a researcher, to the sort of criticism that we had been receiving. The natural inclination is to be defensive and say that there are no problems at all with your research. It is also natural to take the criticisms personally. Sometimes we fell into both traps. But the best answer to criticisms – at least to those that are valid – is to do more research and check out whether or not the criticisms are important.

Even before the first study was published in the *Lancet*, we had started doing a second study.

The one major problem with the first study was that the data on prescribed medicines were taken from different sources for the cases and controls. A new study in which the drug data came from the same sources for cases and controls was needed. We had been aware of this problem from the beginning, but could see no way to solve it. No one else had come up with a solution, either. Since the first study had included deaths between 1981 and 1983, the second study should include deaths from the earlier years of the epidemic, perhaps 1977–81. More than 90 per cent of the deaths occurred outside hospital, and hospital notes relating directly to the deaths were not available. Yet, hospital notes seemed to be the only feasible source of information for the controls, since they would be selected from people who had had a hospital admission for asthma and did not die.

DESIGNING THE STUDY

I kept a graph of the epidemic of asthma deaths on my wall (see Figure 4, p34), and looked at it every day. The answer came to me in December 1988, just before the Department of Health review panel meeting about the first study, when I was working on a completely different problem. I called Richard Beasley straight away to tell him about the idea before I forgot it.

FIGURE 9: Design of the first case-control study

Our first study had included all deaths from asthma (in the 5- to 45-year age-group) in New Zealand. The controls were chosen from patients who had a hospital admission for asthma, so that the controls would have a similar chronic asthma severity to the cases. So the cases were asthma deaths, and the controls were non-fatal asthma hospital admissions which occurred on or about the same dates (see Figure 9).

However, our key findings had occurred in the subgroup of cases and controls who had had a previous hospital admission for asthma during the 12 months before the event we were studying (death for the cases, or admission for the controls). The excess risk from fenoterol was confined to this subgroup of people with chronically severe asthma, and fenoterol did not seem to be particularly dangerous in people with mild asthma. This can be seen in Table 4 (p81), where the data have been split into two rows. The first row contains the data for the cases and controls who had chronically severe asthma (those who had had a hospital admission for asthma in the previous 12 months); the second row contains the data for the cases and controls who did not have chronically severe asthma (those who had not had a hospital admission for asthma in the previous 12 months). It was only in the first group that fenoterol seemed to increase the risk of asthma death. So, if we did another study, we needed only to try to repeat the findings in the first row of the table.

I realised that we could do a further study based on this sub-group of people with chronically severe asthma (see Figure 10).

We could study asthma deaths in people who had had a hospital admission for asthma in the 12 months before they died (the 'index admission'). Then we could choose controls who had had a hospital admission for asthma around the time that the case died, and who had also had a hospital admission for asthma in the previous 12 months (the 'index admission'). For all of the cases and all of the controls, we would then go back to the index admission, and get the information on drug prescribing and chronic asthma severity from the hospital notes.

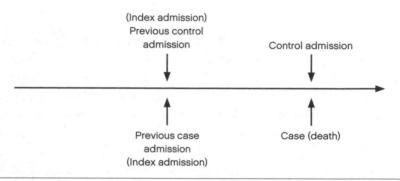

FIGURE 10: Design of the second case-control study

This new design would enable the information on drug therapy to be collected from the same source, hospital notes, for both cases and controls. Furthermore, it would be information on the regular prescribed medication, which had been collected and entered routinely at the time of the admission. It could not have been biased, because at the time of the admission no one knew whether the patient was going to die during the next year. That we would be collecting drug information from hospital admissions which occurred several months (on average) before the death also had its drawbacks, as it would be only a surrogate measure for the medication used around the time of death. However, this problem would apply equally to cases and controls, and we knew that any errors would tend to blur the differences between cases and controls. Thus, if we found a difference between cases and controls (in the proportion prescribed fenoterol), it would be very unlikely to be due to inaccuracies in the drug information.

We knew that this new study was unlikely to be funded by the Medical Research Council or the Asthma Foundation. Once again we would have to do the study in evenings and weekends without any funding. It was better not to let too many people know about it. We had already experienced the denial of access to records from Dunedin Hospital, and we were concerned that further interference might make it impossible to complete the second study.

We started with a list of all asthma deaths in 5- to 45-year olds between 1977 and 1981 (up to July 1981, because the previous study had started from August 1981). We eventually found the hospital notes for 58 patients who had died of asthma during this period, and who had also had a hospital admission for asthma during the previous 12 months. For each of these deaths, the research nurses collecting the data chose at random up to four controls who had had an admission for asthma around the time the case died, and who had also had an earlier admission for asthma in the previous 12 months.

For all of the 58 cases and 227 controls, the research nurses went back to the index admission and abstracted all the information about prescribed medication, and markers of chronic asthma severity. I then copied the information, removing anything which might indicate whether the patient was a case or control – that is, any information to say that the patient had subsequently died. Burgess and Crane then assessed the forms 'blind' – that is, they did not know who were the cases and who were the controls – and recorded what they considered to be the regular prescribed medication for each patient.

TABLE 6: Numbers of cases and controls prescribed fenoterol and salbutamol in the second case-control study

	CASES		CONTROLS			
	Yes	No	Yes	No	Odds ratio	95% CI
Fenoterol	31	27	83	144	2.0	1.1–2.6
Salbutamol	23	35	113	114	0.7	0.4–1.2

The results were as startling as those of the first study.[1] More than half of those who died had been prescribed fenoterol, and the overall relative risk was 2.0 (as shown in Table 6). When the analysis was confined to those with severe asthma, the relative risk for fenoterol increased to 9.8 (see Table 7).

TABLE 7: Numbers of cases and controls prescribed fenoterol in various subgroups of asthma severity

Subgroup	CASES		CONTROLS		Odds ratio	95% CI
	Yes	No	Yes	No		
Admission in past year	22	11	44	86	3.9	1.8–8.5
Oral steroids	12	3	24	35	5.8	1.6–21.0
Admission in past year + oral steroids	11	2	14	25	9.8	2.2–43.4

Thus, where the first study had found that fenoterol increased the risk of death by about 13 times in people with very severe asthma, the second study found that it increased the risk of death by about 10 times in the same group. The overall pattern of findings was very similar to that from the first study, but this time the information had been collected from the same sources for the cases and controls and assessed blind. It was very likely that we were actually underestimating the risks of fenoterol, and it was very unlikely that we could be overestimating them.

NEWCASTLE MEETING
There was going to be a pharmacoepidemiology workshop* in Newcastle, Australia, in June 1989, and this seemed to be an ideal place to pull the second study 'out of the hat'. We submitted an abstract of the findings from the second study, as well as an abstract of the first study, and we were asked by the workshop

---

* Pharmacoepidemiology deals with issues of drug safety and could be more simply termed 'drug safety epidemiology'.

organisers to present the two studies together. It was my first time in Newcastle, so as soon as I got there I went to the Parthenon Milkbar for a milkshake,* before going over to register for the workshop.

The conference involved pharmacoepidemiologists and regulatory authorities from Australia, New Zealand, the United Kingdom and the United States. Also attending were several Boehringer representatives, including Stephan Lanes (who had been flown out from Boston by Boehringer), Ian Mills (the Australian Managing Director of Boehringer, formerly the New Zealand Managing Director) and Daniel Talmont (the Australian Medical Director of Boehringer). Talmont had had the bad luck to move to Boehringer Ingelheim, from a job with another drug company, not long before the fenoterol crisis broke. Of all the pharmaceutical company employees I met during the controversy, he was one of the few who impressed me with his honest approach and his genuine concern for the welfare of patients. He left the company about a year later.

Early on at the meeting, Talmont came to tell us that Ian Mills had 'good news'. We met the two over afternoon tea. They told us that it was unfortunate that our studies had not been supported by the company, and hinted strongly that they could now start giving us lots of funding. I said nothing, finished my tea and left. Julian Crane likes his food, and was concentrating on his tea and cakes and missed the hints completely.

In the days before our talk was due, it became clear that some of those at the conference were not friendly. They had all received the Boehringer publicity packages, and many had become convinced that the first study was 'fatally flawed'. We endured several days of talks, drinks and dinners at which people would bail us up in a corner and ask, 'Why are you doing this?' and what was our response to all the criticisms from the 'international experts'? For example, Michael Hensley, a local respiratory physician, cornered

---

* At the time, Newcastle, Australia, was most famous for the Parthenon Milkbar which featured in the 'Newcastle Song'.

us at drinks, saying, 'I have to talk with you guys.' He told us that he didn't believe our study, and didn't believe the story of the 1960s epidemics. No one had yet heard about the second study, because the conference organisers had forgotten to distribute the abstract. We told everyone to wait until our presentations and said that we would talk with them afterwards.

After our presentations, everything had changed. For the first time, we were able to present our findings to an international audience which, although not particularly sympathetic, still had an open mind on the issue. Crane went through the background to the debate, the development of the fenoterol hypothesis, and the first study. I then discussed the criticisms of the first study, and explained why the non-comparability of the drug information was the only major problem. I then presented the second study, which solved the problem.

When we had finished, John McEwen (head of the Evaluation Support Branch in the Australian Department of Community Services and Health) commented to the person sitting next to him that they had just witnessed an historic moment. This became more apparent as the discussion unfolded. It was introduced by Paul Stolley who said:

> The question of interest is whether a particular asthma medication is associated with asthma deaths in New Zealand . . . they can't use a randomized controlled trial, a cohort study obviously won't work, and so they used the case-control method and you have heard their presentation. The problem about getting unequal histories in the first study I think was clearly the worst problem that they had and the critics correctly focused on that . . . Now this next study . . . deals with this.
>
> So where are we left? We are left with the continuing epidemic in New Zealand, a lot of speculation, a number of dogmatic assertions, but finally we have some investigators who have actually gone and got some data . . . How convincing is it? Well, Henry David Thoreau, a philosopher . . . wrote in his journals during a time when it was common to adulterate milk with river water . . . 'Some circumstantial evidence is stronger

than others, as when for example you find a trout in your milk'. Well I don't know if they have a trout in the milk, but at least they have a minnow, and I think it is very worrisome. What's to be done? Well, they argued about the cause of subacute myelo-opticoneuropathy (SMON) in Japan for ten years, and they did case-control studies and they were all criticized and so on. Finally, the Japanese government banned the halogenated hydroxyquinolines that were implicated and SMON disappeared. I think that regulatory agencies may be too timid. You always have the possibility of doing an experiment in prevention and withdrawing a drug and noting the effects.[2]

The initial discussion related to minor methodological issues. However, the chair, David Christie, a well-known Professor of Occupational Medicine at the University of Newcastle, then suggested:

There seems to be a certain amount of pussy-footing around here. Would someone care to comment on whether this drug should be withdrawn? . . . The reason I ask is that as a simple factory doctor I find myself somewhat baffled because the first question here would be whether there was an acceptable substitute. If this was an industrial chemical the manufacturer would have to say 'No, there is no acceptable substitute' before he was allowed to continue marketing.

Michael Hensley then stood up to speak. I was keenly interested in his opinion because he had been so critical of our first study during the previous few days. He said:

This is a very controversial paper, and as both a respiratory epidemiologist and thoracic physician I approach talking about this with some trepidation. The original *Lancet* article . . . had flaws . . . which really left you in some doubt as to whether there was definitely an association, and . . . whether it was the actual cause of death. The subsequent evidence presented from further studies would have to make you say that . . . there is more likely

141

to be an association than not . . . I think with the evidence that we've seen today, and David's analogy is quite correct, that if you were facing this in an occupational setting, and you had available to the consumer alternative agents, there would be a very strong case to withdraw the agent, if not permanently, certainly temporarily, and see what happens.

Next was David Lawson, Chairman of the United Kingdom Committee for Review of Medicines, who said:

You provoke me when you ask what the regulatory authorities should do . . . It seems to me that certainly I worried a lot when I saw the first case-control study. I worry about precisely the sort of problems you have outlined, and it seemed to me that a relative risk of up to 13 was something which was very difficult to ignore, but we managed (just) to do it in the U.K., since the utilization of this drug in the U.K. is vanishingly small. Now you come up with another study solving what was my great hangup in the original methodology, and your relative risk is 9. It seems to me that there are plenty of other drugs available to treat this . . . I certainly think that if I was in New Zealand I would be pushing hard to get it off the market tomorrow or the day after.

David Christie finished the discussion by concluding that 'the message the troops wish to send to Caesar is that this drug should be withdrawn'.

It was the first time that our studies had had a fair hearing, and a huge change from the flak that we had been getting in New Zealand. We didn't know how to react. The Boehringer representatives were also stunned and had said nothing throughout the discussion. Afterwards, they asked us out to dinner to discuss 'the future'. We declined, but made the concession of having a beer in the bar – the first beer that had ever been bought for me by a drug company. Instead, we ate with some of the other people attending the workshop. I couldn't believe what had happened, and I lay awake for most of the night thinking about it.

At 7 am the complimentary copy of the *Newcastle Herald* was pushed under my door. A reporter had been at the workshop and there was a front-page story on the fenoterol session, quoting the remarks from Lawson, Stolley and Hensley that the drug should be withdrawn.[3] I faxed the report back to New Zealand, where it received extensive coverage.[4] It was time to get out of town, and we headed for the airport. Back in Wellington, the *Sunday Times* carried a story on the controversy by Al Morrison which was very positive about our work and critical of the company and the Asthma Task Force.[5] Stolley and his wife then visited Wellington for a few days, and he appeared on New Zealand television, supporting the second study and reiterating that, as an experiment, fenoterol should be temporarily withdrawn. It was a scientific and psychological turning-point. We all went out and celebrated what had been, for the first time in more than a year, a good week.

CHAPTER 9 **Mixed reactions**

ACCESS TO THE MANUSCRIPT

It was too good to last, and it didn't. The first problem was what to do about the report on the second study, which had only just been submitted to the *Lancet*. The Department of Health would want a copy, and we felt that it should have one, but it would want to pass the manuscript on to Boehringer. The Boehringer machine would then swing into action, and we might again have problems getting the paper published.

The Australian Department of Community Services and Health had also asked for a copy of the paper for consideration by its own regulatory committee. John McEwen (head of the Evaluation Support Branch) stated:

> This Department is prepared to give an undertaking of
> confidentiality in return for a copy of the manuscript . . . It is
> important that you be aware that the Department's undertaking
> cannot be absolute . . . [However], having given the undertaking
> of confidentiality, the Department would maintain its stand . . .
> and a hearing and decision would not occur within a couple of
> months of the request for a copy of the document.[1]

This would give us at least a few months to get the paper accepted for publication. We agreed to these conditions and gave the Australian Department of Community Services and Health a copy on 6 July.

The New Zealand Department of Health was less helpful. Although one departmental official had initially seemed willing to give us an appropriate assurance, negotiations broke down over the details, and it soon became clear that the department would not give us any assurance. Instead, it requested a copy of the manuscript from the university under the Official Information Act.* The first we knew about the department's request was on 5 July, and on 6 July I received a fax from the Registrar of the University of Otago:

> A written request pursuant to Section 12 of the Official Information Act 1982 (as amended) has been received by the University from a responsible officer of the Department of Health for a copy of the paper entitled 'Fenoterol and death from asthma in New Zealand, 1977-1981' . . . The University requests and requires you to forward a copy of this paper to reach me here by 5.00 p.m. tomorrow Friday, 7 July 1989.[2]

Fortunately, the Medical Protection Society had supplied us with a superb lawyer, Hugh Rennie. Meetings with him were becoming a regular event, and we were beginning to regard Rennie as an indispensable member of the group. Time after time we would go downtown to meet him, feeling depressed about a problem which seemed impossible to solve. He would scrawl a note in simple but subtle legal language and the problem would magically vanish. This time the problem was more difficult, however, as we were coming up against our own employers. Furthermore, we did not want an open confrontation with the Department of Health. We felt that the department should see the paper. However, it also had an obligation to protect the paper's confidentiality, and it was the department's refusal to accept that obligation which was the source of the dispute. In the end, we decided that the department should have a copy, despite our concerns, and on Friday, 7 July 1989 we sent it one directly.

The following week, the university, which still did not have a copy of the manuscript, received a further request for one under the Official Information Act, this time from Boehringer Ingelheim. Although we had been forced to give way on the Department of

---

* This applies to universities in New Zealand as they are government institutions.

Health request, this time we were not prepared to hand over the manuscript. In a letter responding to the university's 'request' that we do so, our lawyer stated:

> Our clients offered to make a copy of the paper available to the Department of Health, subject to the Department giving an undertaking that it would not supply the paper in advance to Boehringer Ingelheim. That condition was made necessary by that company's seriously improper use of the previous paper.
>
> The conditions set by our client were proper, and in full compliance with the Official Information Act . . .
>
> The Health Department refused to act in that manner; and made it clear that it would release the paper to Boehringer . . .
>
> Our clients acknowledge the University's full entitlement to receive a copy of their paper. There is however a related duty on the University to act in a way which has the same effect as the undertakings sought from the Health Department.
>
> In light of this letter, our clients trust that the University will agree to take the same position as themselves in relation to release to the Health Department; including the testing of that issue by reference to the Ombudsman (which is a procedure which is available).
>
> Regrettably, if this cannot be done, our clients will be forced to have the issues reviewed by the High Court as between themselves, the University, the Health Department, and Boehringer Ingelheim. It would be unfortunate it that were to prove to be necessary.[3]

The threat to take our employers to the High Court was not going to make us popular. Yet, an important principle was at stake, as well as the publication of the paper itself. It was frustrating that the university was not prepared even to use the usual systems of appeal available under the Act. However, the potential dispute with the university became irrelevant, and the letter was never sent, because the Department of Health gave a copy of the manuscript directly to the company. The department had kept the manuscript for a only week before giving it to the company.

The eagerness of the department to take the manuscript from us and to give it to the company was particularly galling. The Official Information Act had been intended to make information held by government agencies more readily available to the public. Now, a government department was using the Act to take research findings from a university and hand them over to a drug company. Furthermore, the flow of information did not occur in the other direction, and the department declined to give us copies of information supplied to the Medicines Adverse Reactions Committee by the company.

MARC AGAIN

The Department of Health had also contacted Professor David Lawson, Chairman of the United Kingdom Committee for Review of Medicines, and asked him to expand on his comments at the Newcastle meeting. He replied:

> These findings are very difficult to explain short of accepting the probability that fenoterol is causally associated in some way with asthma deaths . . . I said at the conference, that were I to be advising the Minister of Health in New Zealand I would indicate that, in my judgement, the new data provides sufficiently strong evidence to justify removing fenoterol from the market place in the very near future . . . other bronchodilators are available and its loss would not constitute a public health hazard.
>
> New Zealand is one of the few countries where records on all asthma deaths are collected in a most careful manner. This, coupled with the high penetration of the New Zealand market with fenoterol, probably accounts for the association first being demonstrated in New Zealand.
>
> While one could spend many months arguing about the minutiae of the studies presented by Crane and Pearce, the facts are now sufficiently impressive to warrant withdrawal of the drug from the marketplace, at least in my judgement.[4]

However, this advice did not arrive in time for the MARC meeting on 12 July. The committee still did not include an epidemiologist, and its response to the new study was unusual, to say the least. The draft minutes of the meeting were released to the media by the Minister of Health, Helen Clark, a move which probably reflected her exasperation with the committee. The minutes show that MARC agreed that the study implied that fenoterol was associated with an increased risk of death. However, the only action the committee recommended was that animal studies should be done 'in Dunedin' to ascertain the mechanism of death.[5] It seemed that MARC had accepted that the drug was dangerous in practice, but before taking any action the committee wanted to find out if it was dangerous in theory. In response, the Public Health Association said that it was 'mystified by this statement since looking for a mechanism of death presupposes that the association is causal. Furthermore, when there is strong evidence of an association between fenoterol use and human deaths, it is a backward step to be doing animal studies before taking further action.'[6]

The research was to be done by a MARC member, Professor Ralph Edwards, with funding from Boehringer Ingelheim. The Public Health Association subsequently wrote to the Minister of Health to express its concern that:

> A recent feature appearing in the *Dominion* . . . contains
> a statement by Professor Ralph Edwards, member of the
> Medicines Adverse Reactions Committee, that he is receiving
> financial support from Boehringer Ingelheim to carry out
> the animal studies recommended by the Medicines Adverse
> Reactions Committee in July 1989 . . . .
>
> [This] appears to imply that a conflict of interest existed
> within the membership of the Medicines Adverse Reactions
> Committee, when it made the recommendation in July to
> carry out further animal studies of fenoterol and death in
> Dunedin.[7]

Edwards eventually told the *Christchurch Star* that he no longer intended to accept Boehringer funding for his animal studies on fenoterol:

Edwards said because of the incorrect allegations made about his professional status in relation to the proposed animal studies he was certain if he did take up the company funding people would find a way to use this against him.

In an earlier interview Professor Edwards said that both he and Boehringer had agreed they would only do the research if they had an independent peer review to ensure it was the optimum type of research.

He said there was obviously some potential for criticism if Boehringer funded the studies directly without any other review.

'It could have been argued we were in Boehringer's pocket', he said.

Professor Edwards said he saw nothing wrong with the proposal whereby Boehringer was to have made funds available through the Medical Research Council for a number of research studies into asthma.[8]

The Minister of Health was among those who were unhappy with the approach of MARC and the department. Helen Clark later revealed that the department had advised her that MARC had recommended that no additional action was required, and that she should make a statement reiterating the caution previously issued by the department. In fact, close scrutiny of MARC's advice showed that it was silent on the question of additional action and that the department had used considerable licence in interpreting its recommendation in that way. Clark later commented:

It seemed to me that the matter could not rest there. On 19 July I responded saying that the press statement I was issuing would state that the drug would be kept under review, that I understood MARC would be seeking futher information, that I expected it to consult its independent epidemiological advisors and the company, and that I also wanted the name of an epidemiologist to be added to the Committee to be submitted to me urgently. The absence of such expertise on issues like this one seemed to me to have serious consequences.[9]

In a further response to the MARC statement, the Public Health Association called for fenoterol to be withdrawn from the market.[10] The association also wrote to the Minister of Health seeking 'clarification of the pharmaceutical company's role in the process of policy making over fenoterol', and recommending that all members of MARC should divulge to the minister any associations with pharmaceutical companies.

David Clark, a pharmacologist at the Otago Medical School, also called for action in a letter to the Minister of Health:

> I attended the Pharmacoepidemiology Conference in Newcastle, Australia, and heard the evidence presented by Drs Crane and Pearce which linked the beta-agonist drug fenoterol to increased mortality in severe asthmatics. The assessment of the study by eminent epidemiologists from other countries, also at the meeting, confirmed that the study provides extremely strong evidence of the link between fenoterol and mortality.
>
> I have now examined the papers in detail and have spoken to respiratory physicians on the issue, and am convinced that fenoterol should not be available for prescribing by general practitioners . . .
>
> As an academic pharmacologist, completely independent of the Wellington group, and with no commercial interests, I would urge that the Health Department restrict the availability of fenoterol to be available only for use through specialists. In the interests of patients' safety, it is most important that urgent action be taken on this matter.[11]

We too had asked for action to be taken in a letter to the department:

> As you are aware, our research group has recently presented the findings of a further study of fenoterol and asthma deaths . . . [The] transcript is now available and a copy is enclosed . . . [We] note that MARC does not contain an epidemiologist. It

would thus seem appropriate for one or more New Zealand epidemiologists to be co-opted . . . This is too important an issue to be left to persons without the necessary qualifications or experience in this highly technical field.

When our previous findings were published, the Health Department stated that it would monitor international response to our work. Now that this response has been received we believe that appropriate action is urgently required to prevent fenoterol causing further deaths in young New Zealanders.[12]

This was the only time during the debate that we took a more active role and urged that action be taken – although we were careful to avoid saying exactly what we felt that action should be. We were criticised for doing this, although our actions were very mild compared with those of Boehringer and its consultants. We did have subsequent regrets about becoming involved, even temporarily, in issues of policy, as we felt it more appropriate for us to stick to doing the research and to leave policy issues to others. Yet, the situation with MARC had become increasingly untenable, and something had to be done.

DEATH IN NEW GUINEA

Our next encounter with the Boehringer reviewers came at the annual meeting of the Thoracic Society of Australia and New Zealand in Rotorua. We had been allocated only a poster session for the report of our first study, which meant that we would have just two minutes to present the findings. The session was chaired by Sonia Buist, a member of the Boehringer consensus panel, who was in New Zealand as the Boehringer Ingelheim Visiting Fellow. She made our brief presentation even shorter by interrupting us to point out that many people disagreed with our findings. Crane commented that we were already well aware of this.

In the following discussion, Ann Woolcock noted that a severe asthma epidemic had happened before, because severe asthma had suddenly struck the South Fore people in New Guinea. It surfaced during the discussion that the epidemic cases had been treated with fenoterol. In a paper published in a Boehringer-sponsored symposium on 'Aerosol bronchodilator therapy in obstructive airways disease', Woolcock had claimed:

> It is possible for almost all patients to take an aerosol. For example, I have many patients in the Highlands of Papua-New Guinea who not only administer the aerosols effectively themselves but also their peak flow meter readings. These people are alive today because of fenoterol HBr and they know how to take it perfectly.[13]

In fact, 28 per cent of the patients had later died,[14] but no one had made the connection between their treatment and fenoterol.

A SECOND CRITICAL MASS

Walter Spitzer, the chair of the Boehringer consensus panel, was to visit New Zealand, and asked to meet us. I was going to be overseas at the time, but Beasley and Crane arranged to meet with him, together with Doug Wilson (Boehringer Ingelheim, New Zealand) and Victor Hartmann (from the parent company in Germany). Because Beasley and Crane expected the meeting to be difficult, they asked Eru Pomare, as Professor of Medicine at Wellington, to attend as an independent witness. They prepared their own minutes of the meeting, and signed a copy to preserve a record. I won't quote these as they have not been confirmed by the Boehringer participants, but Beasley and Crane certainly felt that they had been threatened and wanted to document this. In particular, they felt that Spitzer had threatened to stop all future publication of our research through an approach to an official committee, on the basis that we were conducting a campaign against Boehringer Ingelheim and using our research for this purpose. I never knew what to make of these minutes and never found out what committee was being referred to. Whatever it was, Beasley and Crane certainly felt intimidated, and when I spoke with Eru Pomare he backed up their description of the meeting.

Next on the agenda was a second Boehringer Ingelheim consensus panel, to be held at the Beverly Hills Wilshire on 28 July 1989. This was the hotel in which the movie *Pretty Woman* was filmed, and is one of the most expensive in Los Angeles. We asked to attend, along with independent reviewers such as Robert Beaglehole, Mark Elwood and David Skegg, but this request was declined. We then contacted those members of the panel for whom we had fax numbers, advising them that:

> We have expressly refused the company permission to transmit the manuscript to any third parties. The manuscript remains entirely confidential and its transmission to you or any other person contravenes the written advice from the authors of the paper to Boehringer Ingelheim Zentrale GmbH . . .
>
> We feel that the interests of asthma patients and their medical advisors would be best served by informed debate of our work in the scientific literature, or at independent scientific meetings . . .
>
> Despite requests by our group for participation in this meeting we have not been invited to attend . . . we do not believe that this is an appropriate manner in which to discuss scientific research. We therefore seek that you decline your invitation to this meeting in the same way that many independent and respected international experts did following their invitation to the New York meeting. We would also ask you to return our . . . confidential manuscript to us without reproduction as an act of good faith.[15]

The Boehringer consultants ignored this request, and the meeting went ahead as planned.

At the time of the meeting, I was teaching in the three-week summer programme of the New England Epidemiology Institute in Boston. The institute was run by Epidemiology Resources Inc (ERI), the consulting firm which employed Stephan Lanes and Alex Walker. As with many areas of epidemiology, there is a substantial overlap between respected academics and private consultants, and it is not easy to draw the line between what is and is not acceptable practice. The summer course was academically of a very high standard, with a respected faculty. It compared favourably with other summer courses, was essentially not-for-profit, and no one seemed particularly concerned that it had the side-effect of lending credibility to the consulting work of the company which organised it.

These ethical dilemmas became more apparent to me when I found myself in the unusual position of teaching at this ERI summer school, while a consultant at ERI, Charles Poole, attended the Beverly Hills meeting about my study, which was being held without my consent and from which I was excluded. I fulfilled my original commitment to teach at the summer school for two years, then resigned from the faculty and never taught there again.

The Beverly Hills meeting included most of those who had attended the previous consensus panel meeting. Stephen Leeder had been invited and checked with us before attending; although we felt that it could compromise his role as chair of the New Zealand Department of Health review panel, it was good to have at least one colleague at the meeting who could let us know what happened.

Jim Hodge, the Director of the New Zealand Medical Research Council, was invited to attend as an observer. His report was complimentary about some aspects of the proceedings, but noted that during the preparation of the final report the participants

> ... were under some pressure from the PR representative ... whose task was ostensibly to translate a technical report into lay language but who clearly held partisan views in favour of Boehringer Ingelheim ... The company representatives prepared the ground for the meeting but stood back from the actual proceedings; participating and commenting only when asked to do so ... Nevertheless their presence could not be ignored and their hospitality was generous, to say the least. The effect of this process on scientific objectivity is always a matter of speculation.[16]

Hodge also noted that the participants wished the report of the meeting to be

> ... devoid of any provocative language and to open with the statements that (i) the second case-control study was better than the previous one and (ii) there remained a possibility that the investigators could be correct but that other hypotheses were

at least as tenable. Only after these two statements were made should the more detailed criticisms of the study be summarized.

However, the report went through several revisions following the meeting, and the final version did not open with these two statements. In fact, it was no more positive than that for the first study, and concluded that

> . . . there are serious methodologic problems with the study design, and that the results can be readily explained by several hypotheses that do not involve an adverse effect of fenoterol . . . The second study avoids only one of the methodologic problems of the first study . . . but has retained others and introduced new methodologic problems . . .
>
> It is expected that the Medical Research Council of New Zealand will oversee a programme for further research in New Zealand with full support and cooperation of public health authorities, of university investigators in that country, and Boehringer Ingelheim.[17]

BOEHRINGER WOOS THE MRC

The reference to the Medical Research Council reflected Boehringer Ingelheim's attempts to obtain its approval for research designed and funded by the company. The Boehringer proposals, which were circulated by the MRC, included the establishment of active surveillance of asthma deaths, and the conduct of a cohort study. The first proposal merely involved repeating the original Asthma Task Force mortality survey, and noted that 'experience with similar studies will be considered an advantage'. The only group with such experience, of course, was the task force. The proposal for the cohort study noted that 'preference will be given to groups which have had experience in such studies, and who have preliminary data from past research of significant patient numbers with severe or life-threatening asthma'. The only group which had collected such data was Harry Rea's group in Auckland.

The MRC at first seemed favourably disposed towards to the research. Its memorandum stated: 'The Council anticipates that following agreement on the scope of the [request for proposals], formal advertisement will be made in due course with the ultimate management and administration of the projects being the direct responsibility of the council.'[18]

In a letter to Hodge, we commented:

> Such an approach could be viewed as Boehringer Ingelheim developing its own research programme which the MRC administers on its behalf, thereby giving it a stamp of respectability. We are concerned that this process would be detrimental to science and the reputation of the MRC.
>
> We believe that if Boehringer Ingelheim wishes to make research funds available to the MRC, then it should do so with absolutely no strings attached. All interested New Zealand asthma researchers (including our group) could then submit grant applications relevant to drug therapy and asthma mortality in New Zealand. For those projects that were approved, the MRC could provide funding from the funds provided by Boehringer Ingelheim . . . Thus, if Boehringer Ingelheim wishes to support research, then it should declare the amount of funds available to the MRC, and the MRC would then decide the way in which the funds would be spent.
>
> A drug company which genuinely believed that there were no problems associated with one of their drugs would have no difficulty in funding the MRC to organise a research programme investigating the safety of their drug . . . If the company is determined to control or limit scientific research into the safety of fenoterol, then the company initiated and funded research must be acknowledged to be precisely that, and the MRC should not be formally involved.[19]

The MRC asked for comment on its proposal from 34 experts, and received 16 replies within 3 weeks. It reported that it had received complimentary comments from only three people, two being members of the Asthma Task Force.[20] A typical comment was that from Dr Don Bandaranayake of the Department of Community Health at the Wellington School of Medicine:

Over the past few months I have been following the unfolding saga in relation to fenoterol and asthma deaths in New Zealand. I have deliberately kept out of the debate as, although I fully supported the findings of Crane et al, I felt that enough hot air had been released by too many people without much forward movement overall.

On finding the real extent of drug company collusion and thinly disguised motives however, I must express great concern about what I feel is an unnecessary involvement of the MRC in a research funding issue which could have far reaching implications. I sincerely believe that the credibility of the MRC could be affected and future research could suffer as a result . . .

To give what amounts to an MRC stamp of approval for a specific drug company to conduct predetermined research is both dangerous and unworthy of an organization which has had an unblemished reputation over the years. What is also of concern is the fact that the proposals are so worded that only very few groups would appear to be eligible to apply for the funds.

I would urge you to reconsider the whole issue and refrain from going down this particular path which is likely to prove treacherous.[21]

The MRC finally decided to withdraw when it was revealed that Boehringer was arranging for direct funding of the research, which would be conducted by a member of the Asthma Task Force. Dr Hodge stated that Rea had been 'attempting to set up an independent mechanism which would bypass council assessment . . . I would wish to question the appropriateness of establishing two separate mechanisms for the funding of substantially similar research projects'.[22]

The MRC's summary of criticisms of the Boehringer proposal started with the comment that 'Boehringer have a poor past record on the fenoterol issue'. The council invited Boehringer instead to make 'unrestricted funds' available for asthma research, but the company did not accept the invitation.

STRIFE IN LJUBLJANA

In September 1989 I was due to teach a cancer epidemiology course in Ljubljana, the capital of what was then the Yugoslavian republic of Slovenia. I wondered whether to make the trip. I was exhausted and felt overloaded, both because of my work and because of the health problems of my parents. My friends had told me that the trip was a good idea, and would provide a break from the pressures of fenoterol. In fact, I was looking forward more to getting a break from the continuing problems with my parents' health.

I had chest pains just before leaving New Zealand, and again while climbing the tower of St Stephen's Cathedral in Vienna, but I went ahead and took the train to Ljubljana. On the third day of the course I stood up to teach, had chest pains again, and the room spun. I made it through the session, but at the coffee break was taken to intensive care, where I was found to have very high blood pressure and was diagnosed with a stress reaction. This was great. I could just lie in hospital, without any responsibilities, and get better. During the next two days my blood pressure stabilised, and soon I was back on my feet.

During the rest of the course I gained an education in the local politics. The course participants, mostly from Yugoslavia, were friendly, but during the coffee breaks I would find myself being hauled into a corner to listen to tirades about the Slavs, the Croats, the Serbs, the Muslims or the Albanians. On my last evening in Ljubljana, the republic's parliament declared that Slovenia had the right to be independent. The next morning I spoke at a cancer conference, but hardly anyone came, because Slovenians were staying off the streets expecting the federal army to arrive. The tanks did not arrive, but federal government planes flew overhead. It was a taste of what was to come when Slovenia declared full independence two years later, and Yugoslavia broke up.

GETTING THE SECOND PAPER PUBLISHED

In the meantime we had been suffering the usual problems of getting the second study published. We had submitted it to the *Lancet*, which turned it down immediately. We then submitted it to the *British Medical Journal*, and on 27 June 1989 I received the standard editorial letter saying that the paper would be reviewed and that this would take six to eight weeks. On 11 July, only two weeks later, I received a further letter saying that the content was too specialised and that we should submit the paper to a specialist journal. This was strange, since the *British Medical Journal* had previously carried many papers on asthma mortality, including some by the Asthma Task Force, as well as others of a more technical nature. In our case, it appeared that the editors had planned to have the paper reviewed, and had then had a change of heart and not waited for the reviews. Certainly, we never received any. Next, the paper was rejected by the *Journal of the American Medical Association* on the ground that fenoterol was not available in the United States.

These rejections were frustrating. It was generally agreed that the paper represented a major contribution to the debate, but after we had made three attempts to get it published no journal had even agreed to send it out for review. Finally, we sent the paper to *Thorax*, edited by Anne Tattersfield and published by the British Medical Association. The paper received favourable reviews and was accepted for publication with minor revisions.

As far as I know, the second paper did not experience any direct interference in publication of the type that had occurred with the first paper. Most of the Boehringer reviewers, with a few exceptions, seemed genuinely concerned about what had happened first time around, and were anxious that it should not be repeated.

However, one problem did arise when Poole, Lanes and Walker from Epidemiology Resources sent a letter for publication in the *New Zealand Medical Journal* criticising the conclusions of the second study. To their credit, they sent me a copy at the same time. However, the letter went beyond what had been presented in the Newcastle abstract and included detailed data from the full manuscript, which was not yet published. In a letter to the editor of the *New Zealand Medical Journal*, I stated:

> I am writing at this stage to express my concern about Poole et al's use of data from our second case-control study which will be published in *Thorax* . . . This manuscript, which was

clearly labelled 'confidential' and 'do not quote' was obtained
by Boehringer Ingelheim without our permission. It has
subsequently been circulated without our permission, and
Poole et al have now quoted this data without our permission
and prior to its publication in *Thorax*. I feel that this is not only
inappropriate, but could jeopardize the paper's publication.
I am particularly sensitive on this point since the activities of
Boehringer Ingelheim and its reviewers very nearly prevented
publication of our previous study in the *Lancet* . . .

I am keen to foster debate on our findings. However, I am
also keen for the usual procedures to be followed in order to
avoid publication of our work being jeopardized.[23]

A heated correspondence followed, in which I stressed that I
was not accusing Poole's group of unethical conduct, but that I
was concerned that their actions might affect the publication of
our work. The Poole letter was eventually rejected by the *New
Zealand Medical Journal*, but appeared in the *Lancet*[24] soon after
our paper was published in *Thorax*.[25] We were not informed of
this, nor offered a right of reply, and our reply was rejected by the
Lancet when it was submitted.

This incident was raised again at an International Epidemiology
Association Symposium on Ethics in Epidemiology in Los Angeles
in August 1990.* Poole spoke before me and commented on my
'volatile state of mind', and my tendency to accuse Boehringer
reviewers of corruption and bias. He also repeated the old allegation
that we had used the Auckland case-control study data without
permission, but Rod Jackson was in the audience and stood up to
set the record straight. In my talk, I tried to ignore what Poole had
said, and concentrated on the scientific side of the story. A report
on the ethics symposium was submitted to the *Lancet* by Professor
John Last of the University of Ottawa.[26] The report was printed
in full, apart from a brief reference to the fenoterol debate, which
was deleted. In justification of this, the editor of the *Lancet* denied

---

* I had at the time no formal involvement in the IEA though I was later elected its president.

that any interference with publication of the original paper had taken place. The *Lancet* seemed too sensitive about the issue even to allow it to be referred to in passing.

MARC MEETS AGAIN

One possible reason for the attempts by the Boehringer reviewers to discredit the second study before it was published was that the New Zealand Department of Health was taking the study findings more seriously this time around. It was beginning to get frustrated with the intransigent stand of MARC, and was beginning to think about ways to bypass this. A report in the *Christchurch Star* on 9 December 1989 stated:

> The department's manager of medicines and benefits, Mr Warren Thompson gives the impression of one who is now worn down and intensely frustrated by the whole affair . . . Warren Thompson claims that, like the Minister of Health, his department has been hamstrung by what he describes as MARC's conservative advice.
>
> 'What I feel about fenoterol or what the department feels is irrelevant. The people who have to be convinced are MARC . . . Except . . . He runs through a hypothetical scenario which could free the department's hands should there be a need to bypass the ministerial advisory committee.
>
> 'Under the Medicines Act we do have the power to approach a drug company and ask them to supply evidence to satisfy certain doubts we might have about the safety of one of their products.
>
> 'If we are not satisfied with that evidence we can refer it to an appropriate committee – now that's a nice little subtlety because it could well be MARC, but it doesn't have to be. We could set up another committee, couldn't we?
>
> 'If that committee then recommended in a certain way we could advise the Minister and take action to have the drug suspended or placed on the restricted list which required special approval for prescription.
>
> 'I'm not saying that that's what we are doing, all the options have yet to be canvassed, but it's a peculiar situation without

precedent and we're trying to ensure that we've done all we can within a limited range of options.'[27]

However, the department had not yet set up a new committee, and the responsibility still rested with MARC. The committee by now included an epidemiologist, David Skegg, and was in a better position to review the epidemiological evidence. MARC was due to meet to consider a review of the second study by Mark Elwood, which had been commissioned by the department. The review had concluded:

Although it still falls short of a totally acceptable standard of scientific certainty, the evidence is strong enough to support clinical and public health decisions, and there do not appear to be strong counterbalancing arguments for the benefits of this particular drug . . . Patients with asthma are managed adequately in many other societies such as the United States, without this drug . . . The balance of the available information is in favour of the causal rather than the confounding hypothesis . . . It is recommended that the drug regulatory authorities should take steps to ensure that the use of fenoterol is minimised.[28]

I decided that it was also time that the general medical research community heard all of the evidence, rather than just the bits they were hearing in the media and from Boehringer Ingelheim. I therefore started giving seminars around the country, going over all the evidence from experimental studies, time trends and the case-control studies. The first was a lunchtime seminar in Auckland, and the Chairman of MARC, Gavin Kellaway, was to be in the audience. Reading the *Dominion* on the plane, I found that my horoscope advised me that 'a lunchtime meeting should be postponed until you know more about the topic you intend to discuss'. I went ahead anyway (starting my talk by reading out my horoscope), and got a good reception, with some discussion about MARC's lack of action on the issue.

MARC met on 6 December 1989. The meeting was described to us by one member as the most unpleasant he had ever attended, but the committee eventually reached a consensus that fenoterol should be removed from the Drug Tariff.* This meant that fenoterol would no longer be available free, except on special application for that small group of patients who had become habituated to it and were not able to switch to another drug. Most drugs were available free in New Zealand at that time, and this would severely limit its use, since other asthma drugs would continue to be available.

THE DEPARTMENT OF HEALTH TAKES ACTION

Thus, the controversy in New Zealand was effectively resolved on 21 December 1989, when it was announced that fenoterol was to be removed from the Drug Tariff. In her press statement, the Minister of Health, Helen Clark, stated that MARC 'agreed that the balance of evidence was in favour of a causal association between fenoterol use and asthma mortality'.[29]

I was in Dunedin that day, at a meeting with Malcolm Sears. Someone came in to tell us about Clark's press release, which had just been issued. I got back to Wellington late in the day. Almost everyone had finished work for the year, and had gone home to prepare for Christmas. A fax from Paul Stolley in Baltimore was lying on my desk. It said:

> What a wonderful Christmas present for your country and
> others at risk from this drug. Your group should receive lots
> of honours (but probably will not) for your cleverness in
> investigating this problem, persistence, and above all, courage
> in following through. I am filled with admiration, but above all
> pleased the government finally did the right thing . . . Again, my
> congratulations to all of you – the fight was lonely, but worth it.[30]

It was the only message we received, and I sat in my office staring at it for about an hour.

---

* The Drug Tariff is issued by the Minister of Health and lists the medicines that may be issued on prescription for free, or with a partial subsidy, with the cost being covered from the government's health budget.

The Department of Health's decision to restrict fenoterol was a turning-point in the debate. The department did not get sued, and it was generally congratulated on having finally taken a stand. An editorial in the Wellington morning newspaper, the *Dominion*, headed 'Fenoterol on the way out', argued:

> The interests of severe asthmatics have finally prevailed. The suspect drug fenoterol will disappear from common use in New Zealand . . . the safety of severe asthmatics has been compromised while other considerations were unfortunately allowed to interfere with the scientific process . . . It is extraordinary that the Asthma Task Force, established in 1978 to research the cause of our high asthma death rate, never seriously examined the possibility that drugs were a factor. The Wellington group used information available to the Task Force years ago.
>
> It is even more extraordinary that the Task Force strenuously opposed the group's findings and resorted to unscientific arguments in an apparent attempt to discredit the study and the researchers.
>
> Meanwhile the manufacturer objected to publication of the original study, criticised the Health Department and blitzed doctors with masses of confounding information . . .
>
> The asthma death epidemic that coincided with the introduction of fenoterol, and doubts cast on the role of drugs in research here and overseas, should have led to research at the time. When that research was finally done, a conservative approach should have been taken, rather than the casual approach that left fenoterol widely available for several months

while under suspicion . . . the fenoterol issue suggests the public cannot be confident its interests are guaranteed.[1]

In the following year, the issue began to be taken more seriously in other countries, and in the medical journals. It was still easy to criticise any single study, but the overall picture – including the 1960s epidemics, the experimental studies in animals and humans, the time trends and our case-control studies – was remarkably coherent. The important thing was to pull all the evidence together and to present that overall picture. I had already given a series of seminars in New Zealand in late 1989, and in 1990 I began to give seminars overseas whenever I had the opportunity. In February 1990 I attended a cancer research meeting at the IARC in Lyon, and this gave me the chance to give seminars also in Milan, Munich, London, Nottingham and Lyon itself.

I've always thought that the purpose of travel is not to have a good time, but to have good stories to tell afterwards. But you can take this principle too far, and I certainly came close on this trip. I arrived in Milan after a long journey, and managed to get to the San Siro stadium to watch AC Milan (with Van Basten) play Napoli (with Maradona) just a few months before the 1990 World Cup. I achieved a first for the San Siro stadium, as I fell asleep during the second half of the match, despite the noise from 80,000 Milan fans. By the time I arrived in Lyon, I had been travelling for a few days, had run out of clean clothes, smelled bad and was exhausted. I was staying in the apartment of a friend who was also attending the IARC meeting. On the first day I just had time have a shower and then to turn the washing-machine on, before setting off to the meeting. I was having trouble concentrating, and my mind was continually wandering over the events of recent months. Finally, I thought about my washing. It would be ready by lunchtime, and I would be able to change my clothes. But had I set the drainage hose in the right position before starting the machine? Making an excuse, I bolted from the meeting. The whole apartment, which was on the third floor, was under an inch of water. The carpet, rugs and mattresses were soaked and the jigsaw of stained wood tiles in the parquet floor had swollen and crunched underfoot as I walked. I panicked. Then I grabbed every towel and sheet in the place, and did my best to soak, sweep and hang things up to dry. I went back to the meeting mid-afternoon and told my friend that I had destroyed her apartment. It was clear that I had outstayed my welcome. I checked into a cheap hotel, spent the night at the Albion, and in the morning headed back to New Zealand.

Most respiratory physicians and pharmacoepidemiologists I met during my 'European tour' had received the Boehringer publicity packages, but few knew the long history of the issue. None had heard our side of the story, except for our brief letter in the *Lancet* replying to some of the criticisms by the Asthma Task Force and the Boehringer consultants. Most respiratory physicians still believed that the 1960s epidemics were originally linked to isoprenaline, but that this had since been disproved. Of even more concern was that many of them also believed that fenoterol was associated with asthma deaths in New Zealand, but that this was because fenoterol was prescribed to more severe asthmatics.

However, some respiratory physicians, and almost all epidemiologists, were beginning to treat the issue seriously, particularly when they had the chance to assess all of the evidence. When they did, the reaction was as overwhelmingly positive as the initial reaction had been overwhelmingly negative.

AUSTRALIA TAKES ACTION

In March 1990, the Australian Department of Community Services and Health followed New Zealand's lead and announced that patients with severe asthma should not use fenoterol. The Australian Drug Evaluation Committee had concluded that 'Fenoterol appears to be associated with an increased risk of death when used in patients with severe asthma . . . The indications for the use of fenoterol [should] be changed to mild to moderate asthma only.'[2]

Boehringer's Australian branch criticised the recommendation, and its medical director, Dr Daniel Talmont, argued that 'Based on the analysis of [the New Zealand] studies, together with other evidence which Boehringer Ingelheim has, the company believes that there has been no change in the safety profile of fenoterol'.[3]

A THIRD STUDY

During 1989 we had also been conducting a third case-control study.[4] This time, the study was funded by the Asthma Foundation and it examined deaths between 1981 and 1987. The basic study design was the same as that of the second study, but we used an additional control group in order to address the remaining criticisms of the previous studies.

The third study found that, whichever control group was used, fenoterol was once again associated with an increased risk of asthma death. However, the alternative control group suggested by the critics from Boehringer actually yielded stronger relative risks than the approach used previously: the overall relative risk for fenoterol was 2.1 using our standard approach, and 2.7 using the approach suggested by the critics.

This third study had become necessary because of the increasingly complicated criticisms raised by Boehringer reviewers. The third study answered these criticisms, and we therefore hoped that the critics might acknowledge this. I presented the study for the first time in August 1990 at a meeting of the International Society for Pharmacoepidemiology near Disneyland in Anaheim, California. Several Boehringer consultants had stressed repeatedly that they had one remaining major criticism of our studies, the choice of control group, and that they would be satisfied when this was addressed. However, in response to the new findings, the Boehringer consultants launched into a familiar attack on the study, and proposed yet another study design which they felt should have been used instead.

The 'official' Boehringer response to our third study came at a third consensus panel meeting in New York in April 1991. The panel was beginning to get depleted, and this time included the old names of Spitzer, Buist, Lanes and Horwitz, as well as the new names of Jean-François Boivin and Samy Suissa (both from Spitzer's department) and Malcolm Sears (who had moved from

New Zealand to McMaster University in Hamilton, Ontario). In addition, Burney, Leeder and Seale sent comments to the meeting but did not attend. The panel's report was much more subdued that the previous two. The panel still found it difficult to comment positively on the study, although the report did acknowledge that the study showed that 'control selection bias' did not explain our findings (this was the hypothesis repeatedly raised by Lanes and Poole). However, the rest of the report largely involved repetition of criticisms made of the previous studies, which had already been addressed.

THE SEARS STUDY

The real surprise in 1990 was Malcolm Sears's study on regular use of fenoterol. For many years, it had been suggested that regular use of beta agonists might make asthma worse rather than improve it.[5] Beta agonists were originally intended to relieve the symptoms only, and not the underlying inflammation. However, around 1979, some asthma specialists began to hypothesise that beta agonists should be used regularly (four times a day) rather than just for relief of symptoms.[6] This 'feeling' was based entirely on anecdotal clinical experience, and there was very little real evidence to back it up. However, it gave pharmaceutical companies the rationale to begin marketing beta agonists for regular use, and their sales soared.

During the 1980s, asthma mortality increased slowly in a number of countries, including the United States and Australia, although none other than New Zealand had an epidemic.[7] Some researchers felt that this slow increase in mortality could be due to the regular use of beta agonists making asthma worse (this is the *severity hypothesis* discussed back in Chapter 1).

Malcolm Sears had been sceptical about this hypothesis, and had criticised it several times in correspondence in medical journals. However, to his credit, he set up a study to test it. Using volunteers, he compared their asthma symptoms during the period when they were using fenoterol regularly (in addition to using it for symptomatic relief) and when they were using the drug only 'as required' for relief of symptoms. Sears used fenoterol for the study, because he was already using it in most of his patients; his study began before our studies first raised concerns about the drug's safety.

The results were clear cut and surprising: the patients did much worse when they were using fenoterol regularly. It was not clear, however, whether the problems had occurred because the drug being used was fenoterol, or whether they would have occurred also if a different beta agonist had been used. Sears did not believe that fenoterol was markedly different from the other beta agonists, and he described the study as involving 'regular use of beta agonists' rather than 'regular use of fenoterol'. Other critics considered that the problems might apply to all beta agonists, but be particularly severe with fenoterol. Sears subsequently concluded that fenoterol might be particularly dangerous because it was marketed in too high a dose.[8]

Sears's study was published in the *Lancet* on 8 December 1990.[9] In the same issue was a laboratory study from Ann Tattersfield's group at the University of Nottingham, which confirmed our previous findings that fenoterol had greater cardiac side-effects than other commonly used beta agonists, such as salbutamol.[10] In an accompanying editorial, the *Lancet* commented on the two studies, discussing other evidence of the hazards of fenoterol and other beta agonists, and calling for the use of beta agonists in asthma to be redefined urgently.[11]

SASKATCHEWAN STUDY

The strongest evidence in support of our findings came in 1991 from a study done in Saskatchewan by Walter Spitzer and several other members of the Boehringer consensus panel.[12] This study had been set up by Spitzer, and funded by Boehringer, in response to our New Zealand case-control studies. It involved 44 asthma deaths in Saskatchewan between 1980 and 1987, and 233 controls.

The findings were even stronger than those in New Zealand. Nearly half of the patients who died had been prescribed fenoterol, compared with 16 per cent of the controls. Patients prescribed fenoterol were five times as likely to die as patients prescribed other beta agonists, whereas our overall relative risks in New Zealand had been only about twofold.

However, the basic results regarding fenoterol, which confirmed the New Zealand findings, were not mentioned in an abstract which was circulated widely before the publication of the paper. The abstract did not even mention the fenoterol hypothesis, which was the main reason for doing the study. Instead, it focused on the possibility that there could be problems with all beta agonists at high doses (a 'class effect'). In the draft report, the fact that the risk of death was higher with fenoterol than with other beta agonists was mentioned only briefly, and was then dismissed on the grounds that this was merely a dose effect and that the risk would have been less if fenoterol had been marketed in a lower dose. This assertion has been contradicted by laboratory studies which show greater cardiac side-effects with fenoterol even at lower doses, and by epidemiological evidence that asthma mortality epidemics have occurred only with isoprenaline forte and fenoterol, and not with other beta agonists. In particular, there was no epidemic when salbutamol was introduced, despite the massive sales.[13] More importantly, the assertion ignores the fact that fenoterol was marketed in a high dose and that the lower-dose preparation was not available at that time.

The abstract was presented at many scientific meetings, and was reported in the *New York Times*, the *Lancet* and the *New Scientist* before publication of the full paper.[14] This led to confusion about the safety of beta agonists in general, and obscured the fact that the study had confirmed the New Zealand findings on fenoterol. The draft of the paper, which was given to regulatory authorities before publication, also concentrated on the possible class effect, and even claimed that 'the results from this Saskatchewan case-control study directly contradict the Wellington findings'.[15] Fortunately, this claim did not appear in the version that was finally published.

Once the Saskatchewan study had been published, a number of problems with its analysis and interpretation became apparent. In particular, there was no evidence in the initial analyses of an increased risk with beta agonists other than fenoterol, and the possibility of a class effect emerged only in additional analyses which involved questionable methods and assumptions. Even in these additional analyses, fenoterol – in the dose in which it was marketed – consistently showed greater risks than the other beta agonists, and the data were very consistent with the New Zealand findings.[16]

The accompanying editorial in the *New England Journal of Medicine* concluded that there was enough doubt about the safety of fenoterol in particular to avoid it altogether.[17] Michael Hensley, a member of the advisory panel for the study, subsequently published a letter in the *Medical Journal of Australia* which noted that the Saskatchewan study supported the New Zealand findings on fenoterol, and concluded that 'the weight of evidence and availability of alternative medication should lead to a recommendation that fenoterol not be used to treat asthma'.[18] Nevertheless, the extensive pre-publication publicity about the paper had had a considerable impact in shaping the acceptance of the authors' interpretation of the findings, and many respiratory physicians apparently still believe that the Saskatchewan study in some way contradicted the New Zealand findings. Those who do believe that there was a problem with fenoterol generally think that it was just that the dose was too high and that the problem has been 'solved' by lowering the dose. Very few believe that the drug is actually different from other beta agonists.

Despite the misrepresentation of the findings, after the publication of the Saskatchewan study it became generally accepted, at least by epidemiologists, that fenoterol (in the high-dose form in which it was marketed) was causing asthma deaths, even if no one could agree on the likely mechanism. As John McKinlay, Professor of Sociology and Medicine at Boston University, wrote to me:

> The [Saskatchewan study] more or less vindicates you and your colleagues.
>   Of course, what you should not miss is the fact that appearance in the hallowed New England Journal of Medicine – that organ of conservative medicine which serves a legitimating

function . . . puts the whole issue to rest. When you write about the diffusion of a technology you have to remember that the NEJM is the anointing agent on high. If you do not subscribe to the NEJM you are not one of the chosen few. If you do not publish in it, you are a member of a sect (not one of the major denomination). If you don't read it you are out of touch. And if a research finding does not appear in it, it must be considered heretical or suspect. If you ever write on the history of the whole fenoterol controversy, may I strongly suggest to you that the appearance of this final article in the NEJM represents the final benediction.[19]

Thus, the Saskatchewan study, despite the protestations of its authors, essentially laid the fenoterol debate to rest. It is now clear that fenoterol caused the epidemic of deaths. It is still not clear whether this was just because the dose was too high, or whether the drug was different from the other beta agonists in some way. Although only the New Zealand, Australian and Japanese medical authorities have formally taken action against fenoterol,[20] the company forestalled action in other countries (and implicitly recognised that the dose had been too high) by halving the dose to 100 µg per puff throughout the world.

FOOD AND DRUG ADMINISTRATION

As a result of the Saskatchewan study, and our New Zealand studies, the United States FDA held hearings into the safety of beta agonists in Washington, DC on 11–12 December 1991.[21] I was asked by a drug company to testify, but I declined, as I was not prepared to appear as a representative of a drug company. Eventually, Richard Beasley and I were invited by the FDA itself as panel members.

In some ways the arguments at the meeting resembled those we had been having over coffee at meetings in New Zealand for the previous few years. Beasley, Robin Taylor (a colleague of Malcolm Sears in Dunedin), and I were special guests on the panel, as were Spitzer and Buist. Doug Wilson testified on behalf of the American branch of Boehringer Ingelheim, and Paul Stolley testified on behalf of another drug company – he was forced to do this because he had not been invited onto the panel, but he stressed that he was presenting his own views and not those of the company.

The proceedings were held in public, unlike the proceedings of New Zealand committees such as MARC. They were video-taped and recorded for the congressional record, and there was a large public audience. This included members of the press and drug company representatives, as well as other people whom I did not recognise. I was puzzled as to why they would go to the telephone during the coffee breaks. Only later did I find out that they were financial analysts working for stockbrokers. The drug companies' share prices were rising and falling as we argued about study design.

The first day saw heated arguments between Spitzer and me. We were seated next to each other on the panel, but the next morning I found that other members of the panel had moved our seats so that we were no longer sitting together. I was asked to answer an 'anonymous' allegation that my presentation the previous day was not independent and had been made on behalf of a drug company. I therefore had to make a public statement on the funding of my studies and to reiterate that they were funded by independent bodies, including the Health Research Council and the Asthma Foundation, and that I did not accept drug company funding for them, or any of my other studies.

The highlight of the day was a hilarious presentation by Paul Stolley which followed one on the importance of house dust mites (and particularly their faeces) as a trigger for asthma attacks. Stolley explained how the 1960s epidemics had been due to isoprenaline forte, but that history had been rewritten and doctors were still searching for alternative explanations. He said that he had been collecting these alternative theories, some of which were very inventive, and that his favourite one, which he intended to submit for publication under a pseudonym, was that the 1960s epidemics had been caused by outbreaks of diarrhoea in house dust mites in six countries, but not in others. Stolley finished by observing:

The United States has been spared two epidemics of asthma mortality, because . . . the FDA has not licensed either fenoterol or isoprenaline in its forte concentration. Now those people who believe that the FDA imposes an unnecessary burden on the industry because of their rigorous premarketing scrutiny of drugs, might rethink their position, because the epidemics occurred in those countries where drug regulatory agencies are unfortunately weaker than ours.[22]

At the end of the meeting Robin Taylor gave a farewell speech in which he said that the New Zealanders had stirred things up 'and now we are leaving', and then we all rushed for the airport.

The trip home was almost as exciting as the FDA meeting itself. At Los Angeles airport we ran between the terminals and just made the Air New Zealand flight as it closed. As usual, I had an economy-class ticket but Richard Beasley had a business-class ticket – somehow he always did. They had put us both into economy by mistake, so he took the tickets back to the gate and got us both put into business class. The pilot aborted the takeoff because of a warning light, and an official came onto the plane to move me back to economy, but by then I was too exhausted to move and eventually he gave up trying. After another aborted takeoff we eventually made it into the air as far as Hawaii, where the plane gave up the ghost. Our luggage vanished, and we spent the day trying to sleep in the sun and heat. Eventually we made it home a day late. When we got back to New Zealand it was the end of the year, and we had the usual Christmas Eve beers in Richard Beasley's office. This year they did not taste so bad.

THE END OF THE EPIDEMIC

The beer tasted even better the following year with the news that the death rate in New Zealand had dropped sharply after the publication of our first study.[23]* Sales in New Zealand fell from a

---

* The mortality statistics are always published a couple of years late because of delays in recording deaths that require a coroner's hearing. Although the death rate dropped sharply after our first paper was published in April 1989, we knew about it only a couple of years later.

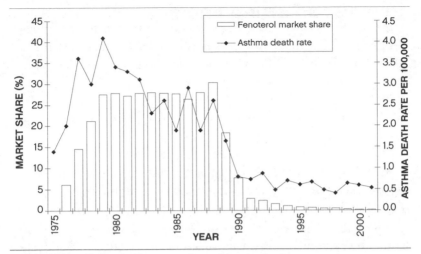

FIGURE 11: New Zealand asthma mortality and fenoterol sales

30 per cent market share in 1988 to less than 3 per cent in 1991 (see Figure 11). In the first half of 1989, the death rate in the 5- to 34-year age-group was 2.2 per 100,000, which was similar to that in the previous 6 years. Following the publication of our first study, and the resulting publicity and warnings in mid-1989, the death rate fell to 1.1 in the second half of 1989. In 1990, it fell further to 0.8, and it has stayed at this level ever since.

They would keep arguing, but I didn't care. The epidemic was over.

A new scientific truth does not triumph by convincing its opponents and making them see the light, but rather because its opponents eventually die, and a new generation grows up that is familiar with it. – MAX PLANCK[1]

Could the second New Zealand mortality epidemic have been avoided if knowledge gained from studies of the 1960s epidemics had been properly heeded? It is difficult to respond appropriately to such questions during the height of a bitter controversy. It takes many years to get these issues into perspective, to forget about the criticisms and the personalities and to be able to look at the big picture – which is why I waited more than 15 years to finish this book. With the benefit of hindsight, it may be useful to look back on the fenoterol story in order to see what lessons may be learned which might help avoid such problems in the future.

EPIDEMIOLOGY

Why is epidemiological research so often full of controversy? The main reason, perhaps, is that epidemiology deals with hazardous drugs and chemicals for which it is unethical and impossible to do a randomised trial. Thus, it is impossible to do a perfect study, and epidemiologists must learn to review all of the available evidence rather than attempting to reach a decision on the basis of a single study.

Epidemiologists have become experienced at coping with this imperfect situation, and at trying to understand the different biases that can occur in epidemiological studies and the best ways of avoiding or minimising them. However, these advances in epidemiological methods are rarely acknowledged by critics from other disciplines. For example, the studies linking smoking with lung cancer were bitterly criticised by 'conventional' researchers who were not willing to accept evidence from studies where the exposure had not been randomised. However, the preliminary evidence that smoking caused lung cancer[2] was eventually supported by hundreds more studies in other countries.[3]

A further reason that epidemiology is full of controversy is the high stakes involved for companies which have produced a drug or chemical that is suspected of causing disease or death. An article in the *Epidemiology Monitor*,[4] published at the time of the fenoterol controversy, raised issues about the limits that should be set on a company's efforts to defend its products when they have been implicated in serious side-effects. Clearly, a company has a right to argue against what it believes are weak data or incorrect conclusions, and it is not in the interests of society or the company to withdraw a drug which has been wrongly accused. However, equally clearly, a company has a moral obligation to seek the truth of the matter when obtaining advice from consultants, rather than just preparing the 'case for the defence'. The latter usually occurs, even if the hired consultants are relatively neutral in the dispute and merely sit on an 'expert panel'. This is not to imply that deliberate corruption occurs – that seems to be very rare. However, a company which intends to prepare the 'case for the defence' may seek out academics who (usually because of sincerely held beliefs) have been very critical of similar studies in the past. As Paul Stolley observed:

> If you hire somebody to look at a paper which has a new
> discovery, and there have been obvious difficulties in
> doing the study, and this guy doesn't believe that DES
> [diethystilboestrol] causes cancer of the vagina, that there's not
> enough evidence to implicate tampons with toxic shock, that
> there's not enough evidence to believe estrogens are related to
> uterine cancer, has even questioned the cigarette smoking/lung
> cancer association. If that's the guy you hire, you don't have
> to be a genius to figure out where he's going to come down on
> this issue.[5]

Thus, the shaping of the 'case for the defence' usually involves 'selection' rather than 'coercion' of experts, although subtle forms of influence may also occur. This is typified by remarks from an American lawyer, John C Shepherd of St Louis, who was President of the American Bar Association in 1984–85:

> The first thing you need to get along with your expert witness is money. But the hiring and successful use of an expert may not be that easy – a lot of good experts are rich. Although you will eventually be talking about money with your expert, it is wiser to begin on another tack. Tell your expert how justice will be served if he will testify on your side of the case. Remind him how the unfortunate situation in our courts today can be improved if we have people of his caliber to help in the administration of justice. That ploy will impress even the rich expert.[6]

The response of consultants may also depend heavily on what question is posed by the company. One possible question would be, 'Is there any chance that the data are right?' To which the answer is always 'Yes'. Most commonly, however, the question posed is, 'Is there any chance that the data are wrong?' To which the answer is also always 'Yes'. This is, perhaps, the question which is most appropriate in the scientific context, where the emphasis is on scientific criticism and debate – we should not accept that something is proven just because an association has been found in one study. However, in the context of public health decision-making, the most appropriate question is, 'On balance, what conclusion is most likely to be true from the data?'* Clearly, quite different reviews will be produced depending on which question is asked.

Perhaps the most legendary epidemiological consultant was the late Alvan Feinstein at Yale University, who features in the

---

* This question is related to 'the precautionary principle', which argues that scientific uncertainty should not be used as a reason to postpone preventive measures to protect human health. It is being increasingly widely used in public health decision-making.

current story. Before his death in 2001, Feinstein disputed most of the major epidemiological findings in recent decades, including the established causal associations between smoking and lung cancer, between oral contraceptives and thrombo-embolism, between diethylstilbestrol and vaginal cancer, between aspirin and Reye's syndrome, between tampon use and toxic shock syndrome, and between estrogens and endometrial cancer.[7] In each instance, these controversies were eventually resolved with the vindication of the original studies, but the debates often lasted for many years, and the necessary safety warnings and regulatory procedures were therefore delayed.[8]

It should be stressed that criticism plays an important role in science, and even very biased critics may occasionally make important points. However, an overemphasis on criticism can lead to the dismissal of almost any scientific study as being 'fatally flawed'. As Stolley writes:

> A . . . distressing development has been the attitude of some self-proclaimed pharmacoepidemiologists that their job is to attack competent studies as consultants to drug companies, who pay them handsomely and even award grants to their research unit as a form of reward. Often the sponsorship of these 'disinterested reviews' is not clearly stated. The most pernicious of these articles are characterized by an unwillingness to focus on the totality of the evidence and a concentration on real and imagined flaws that could not possibly account for strong associations.[9]

At a conference on ethics in epidemiological research, I once presented a satirical set of guidelines for a 'corporate epidemiologist' who is asked to review a study:

1. Consider only the specific study that you have been asked to review. Don't consider supporting evidence from other epidemiologic or experimental studies.
2. There are three possible questions you could consider: (1) is there any chance that the study findings are right? (2) is

there any chance that the study findings are wrong? (3) what is the balance of evidence? Restrict yourself to the second question.

3. Prepare a list of possible biases. Do not comment on the likely direction or magnitude of the biases. Conclude that there are many 'fatal flaws' in the study and it is therefore uninterpretable.
4. Decline to comment directly on policy, but insist that further studies must be undertaken which avoid the biases identified in step 3.
5. Go back to step 1.

The selection by a pharmaceutical company of a few scientists who follow this procedure, and are hypercritical of others' work, can result in massive pressure on public-health decision-makers. This pressure is particularly effective because it seems to come from independent scientists – it would not be taken so seriously if it came directly from the company. In this sense, the company's consultants have the privilege of acting as 'lawyers for the defence' while maintaining the image of being an 'independent jury'.

This type of reviewing also poses problems for a journal which is considering for publication a paper that implicates a particular drug in serious side-effects. Ideally, a good journal should stand by its own reviewing process, and should not consider any unsolicited reviews. However, as the *Epidemiology Monitor* points out, if reviews commissioned by a company are sent to a journal, 'the journal is placed in a difficult position because it may feel that it cannot prudently ignore the criticisms, yet they may not have been obtained during the normal process of peer review'.[10]

ETHICS

Another reason that it is becoming harder to do epidemiological research into drug safety is the increasingly difficult situation with ethical approval. We did the fenoterol studies just as the Cartwright Inquiry was leading to a reorganisation of the system of ethical approval in New Zealand.[11] The Cartwright Inquiry arose out of evidence that unethical research was being conducted without informed consent by Professor Herbert Green at National Women's Hospital in Auckland.[12] The research involved withholding conventional treatment from patients with carcinoma *in situ* of the cervix in order to study the natural course of the disease; about 40 of the women eventually developed invasive cancer.[13] The story was made public as a result of the work of some of Professor Greene's colleagues, who were able to study the patient records and to show that the women involved in this unethical research (none of them knew that it was happening) were 25 times more likely to develop invasive cancer than women receiving more conventional 'treatment'.[14]

The story was known in the medical community for years before it became public when Sandra Coney and Phillida Bunkle published an article about it in the monthly magazine *Metro* in June 1987.[15] I first heard about it a couple of months later in the women's toilets in a bar in Helsinki. I had just spent two years working in the United States and was taking the long way home via Europe, including a meeting of the IEA in Helsinki. I went out for a beer with the other New Zealanders at the meeting, keen to hear the news from home. We went to the bar on the top floor of the tallest building in Helsinki, and I was surprised to be told that 'There's a really good view from the women's toilets'. Someone covered the door while a few of us went to have a look, and it was true – there was a magnificent panorama over all of Helsinki. On the way out, Ruth Bonita (then an epidemiologist at the Auckland Medical School, but now a senior official with the World Health Organisation) started telling me about this article that had just been published that was going to change health research and clinical practice in New Zealand for ever. It was expected that the exposure of the story would be a huge boost for epidemiology, because it was an epidemiological study of hospital patient records that exposed what had been happening, and that women were dying from cervical cancer as a result.

The reality was quite different. The Cartwright Inquiry led to a major overhaul of the system of ethical review in New Zealand, but had at best a small impact on clinical research, and almost no impact on drug company funding of research. Instead, a strong emphasis on privacy led to massive restrictions on epidemiological research – a classic case of silencing the whistle-blowers. Nowadays it would be very difficult for Professor Green's colleagues to 'blow the whistle', because they would have great difficulty in getting access to the patient records to show that an excess number of deaths from cervical cancer was occurring.[16] Similarly, we would have great problems nowadays in obtaining access to the information we needed from asthma hospital admissions in order to do the case-control studies that showed that fenoterol was causing the epidemic of asthma deaths. So there is a danger that, when the next 'fenoterol story' happens, it might never see the light of day, because it may be impossible to get ethical approval to study it.*

FUNDERS

Although the problems that occurred with the fenoterol studies are not unusual, the situation was made particularly difficult by the existence of the Medical Research Council's Asthma Task Force. It is hard to avoid the conclusion that the creation of the task force in 1978 delayed the discovery of the cause of the New Zealand epidemic by many years, and stifled the research that really needed to be done to establish that cause. Furthermore, it strongly affected the response to our research by the MRC, which came dangerously close to supporting the activities of Boehringer Ingelheim in trying to 'manage' further research into the safety of fenoterol.

Why did the MRC create the Asthma Task Force? Health research funders like to appear to be doing something. They cannot easily influence the kinds of applications they receive in

---

* The situation may improve with the recent publication of the report of the National Ethics Advisory Committee, of which I was a member. This reviewed the processes of ethical review of health research in New Zealand, particularly some of the difficulties of conducting epidemiological research, and recommended improvements to the system. (A. Moore, M. Ardagh, D. Bramley, A. Bray, et al., 'Review of the current processes for ethical review of health and disability research in New Zealand: report to the Minister of Health'. Wellington: National Ethics Advisory Committee, 2004). However, the impact of the changes resulting from this report is still unclear.

their regular funding rounds, where researchers can think of their own hypotheses and apply for funding directly for 'investigator-initiated' research. So it is easier for them to appear to be 'doing something' by creating a task force, or some other body which sounds official and which encourages or requires researchers to 'all work together'. More often than not, the creation of such research monopolies soaks up all of the available research funding in a particular area, and makes it very difficult for other researchers with alternative hypotheses to get funded to do their research. These problems are particularly acute in small countries like New Zealand where there are few sources of alternative funding for researchers who have hypotheses that don't fit the current establishment thinking.

Perhaps the most famous example of the dangers of such 'research planning' is the story of the discovery of the double-helix structure of DNA by Francis Crick and James Watson, who drew on their knowledge of work by Rosalind Franklin and Maurice Wilkins.[17] The British Medical Research Council several times interfered in the research by trying to decree who had their mandate to work on DNA and who didn't. Watson and Crick went ahead anyway, and the rest is history. In the case of DNA, of course, someone else would have soon discovered the double-helix structure if Watson and Crick had not done so. However, in many other cases, the creation of a government-funded research monopoly in a particular area such as asthma means that some questions may never get asked, because they are not priorities, or are embarrassing for the government agencies or 'opinion leaders' in the particular field of medicine.

So what is the solution? To recognise the key role of independent university-based research, and to ensure that substantial funding is always available for investigator-initiated projects. Such research is often criticised for being 'ivory tower' or 'out of touch with current priorities' or 'lacking support from end-users'. This is usually just another way of saying that the authorities do not see the research as a priority because it might ask uncomfortable questions, or might identify problems that the authorities have to do something about. While there is a role for targeted funding for task forces, and centres of research excellence, these should never be encouraged or allowed to have a monopoly or to stifle intellectual competition and debate. There must always be an important role for the outsider, the devil's advocate, the heretic.

DRUG COMPANIES

The difficulties of having evidence accepted that medicines in common use may be worthless or even dangerous, are strengthened by the close relationship between doctors and drug companies.[18] At the time of the fenoterol studies, drug companies in New Zealand spent about $25,000 per general practitioner each year on advertising and promotion.[19] One GP reported receiving 3 kg of mailed advertisements in the first four weeks of 1989 (i.e. just before the fenoterol controversy broke). During the same year, the New Zealand Department of Health provided two issues each of the *Clinical Services Letter* and *Therapeutic Notes*, a total of 30 pages, to provide information on the 8000 different medicines (containing 2000 different active substances) on the New Zealand market. Because medicines are subsidised in New Zealand, the money that Boehringer Ingelheim used to influence doctors actually came from the Department of Health budget and therefore from taxes, but the department did not have enough resources to counter the activities of the company even if it had wanted to.

The drug company 'sponsorship' of doctors includes not only direct advertising, but also countless free gifts, ranging from pens and prescription pads to free meals, bottles of wine, cellular phones, holidays and educational seminars in exotic locations. Back in the 1970s, the director of the United States FDA Bureau of Drugs stated:

> The growing influence of the pharmaceutical industry on medical education is a long-term threat to the integrity of the medical profession . . . It is inevitable that the educational materials produced by and for an industry with an interest in increasing sales of drugs will, on balance, be biased in a direction intended to promote drug use.[20]

Most doctors are unwilling to reveal the extent of their gifts from the pharmaceutical industry. Privately, many of them argue that they are unaffected by such gifts and that it would be foolish (rather than moral) to reject them. However, as a 1989 article in the *Journal of the American Medical Association* pointed out:

> Inherent in the relationship is an obligation to respond to the gift; this obligation may influence the physician's decisions with respect to patient care or possibly even erode the physician's character . . . The companies are, of course, motivated by profit, not altruism . . . Also, the fact that many physicians are not seasoned business people who are aware of subtle but compelling sales techniques probably contributes to the success of these marketing tactics.[21]

I have seen this process in action at meetings of the Thoracic Society of Australia and New Zealand – the main meeting that most respiratory physicians attend each year. As well as the lavish trade displays and 'freebies', the companies competed to take influential doctors out to dinner and show them a good time. Company representatives got particularly upset if a researcher they had been funding went out to dinner with representatives from another company.

What was particularly noticeable during the fenoterol saga was that many respiratory physicians and general practitioners privately expressed sympathy for the company, and concern that the New Zealand branch might have to close down if the drug were withdrawn. Only rarely did I hear a respiratory physician or general practitioner express concern for the patients who were using the drug. It was clear that fenoterol was no better than other available drugs, and might be much worse, and that safer alternatives were readily available, but many doctors were keeping their patients on fenoterol nonetheless.

CLINICIANS
However, even without the pharmaceutical company influences, it is unlikely that the evidence about the dangers of fenoterol

185

would have been easily accepted by clinicians. Almost all major discoveries in medicine have been met with bitter criticism and lasting controversy as leaders of the medical establishment have found their traditional ideas and world-view threatened by new ideas. In many ways, this parallels the situation in other fields of science, but there are some features of medical research which can make the struggle between new and old ideas particularly difficult and bitter.

One reason is that most methods of treatment have grown from a tradition of summarising the clinical experience of individual doctors. This often amounts to no more than the 'gut feeling' of an eminent doctor based on their experience of treating individual patients. Although such practical experience can provide great insight, it can also be misleading, because every patient is different and the experience of an individual doctor is often too limited to draw general conclusions from. For example, the average general practitioner would have only one asthma death in their practice every 10 years. If a new drug is introduced that doubles the risk of dying from asthma, the average GP would have two deaths in their practice every 10 years rather than one – they would probably not even notice that the death rate had changed, and would certainly not notice that deaths were more common in people using the new drug.

But that is not the only problem. Even when established treatments have been tested in randomised trials and clearly shown to be ineffective or dangerous, doctors have been reluctant to accept the evidence. There are many such examples from the history of medicine, but one of the most famous is that of Ignaz Semmelweiss, who discovered that puerperal fever ('childbed fever') could be prevented by requiring doctors to wash their hands in chlorinated lime before entering the ward, and between each examination.[22] His findings were not accepted by most of his medical colleagues, who took offence at the implication that they had been killing their patients with their unhygienic practices. Semmelweiss was denied promotion, eventually became mentally ill and died in a mental hospital of 'childbed fever' contracted when he cut himself during his last obstetric operation. It was another 30 years before his ideas were revived and began to gain acceptance.

More recently, Sir Archie Cochrane, in a BBC documentary, described the reluctance of doctors to accept the findings of a randomised trial of coronary care units compared with home care for patients who had suffered a heart attack. The trial eventually found that patients cared for at home had a slightly lower death rate in the first year after their heart attack than the patients cared for in hospital. Cochrane had great difficulty getting permission to conduct the trial, which was considered unethical by coronary care doctors. Eventually it was allowed to proceed, provided that the progress was regularly reviewed. The initial results showed a slightly smaller number of deaths in the group with home care (six deaths) than in the group with hospital care (eight deaths):

> Just for fun I reversed the table, showing more deaths at home than in hospital . . . I showed it to some of the consultants before the meeting and there was an absolute uproar . . . 'that trial is unethical, it must be stopped' . . . so I let them go on a bit and blow their tops, and when they were calming down a bit I apologized that I had shown them the wrong table. I then showed them the correct table, and said didn't they think it was unethical to continue with coronary care units, but I was unable to convince them . . . it does make the point that there is an enormous amount of emotion about coronary care units.[23]

Despite the reluctance of many doctors to accept the evidence about fenoterol, it is also important to stress that there are heroes as well as villains in the fenoterol story, and that most of the heroes (as well as some of the villains) were medical doctors. Doctors are probably no better or worse than anyone else in their reluctance to accept bad news. It is just that, because they are doctors, we expect that they should have the patient's best interests at heart, and it is a shock to realise that this is not always the case. They are simply humans required to do a superhuman job. Having worked for 20 years in a medical school, and 10 of those right next to the intensive care unit, I can't help but admire the courage and professionalism that most doctors need just to get through each day's work. I've sat in the cafeteria at lunchtime with doctors who have just had to tell someone they were going to die, or who have just had a young patient die, or who have been threatened with legal action, bad publicity or physical assault by patients or grieving relatives. It's not an easy job, and people who do it have a shared bond. That's why it takes a particular kind of courage to 'swim against the tide' and to tell your friends and colleagues that their treatment may be killing their patients. It is also easy to slip into believing that 'doctors can do anything' and that a doctor who is good at treating patients will also be good at research. Many of them are, but a brilliant clinician can also be a terrible researcher, particularly when they step outside their field of expertise and get involved in epidemiology – like an opening bowler who bats at number 11.

MINISTRY OF HEALTH

There are two key questions about drug safety for what is now the Ministry of Health: 'under what conditions should a drug be licensed?' and 'under what conditions should this licence be restricted or revoked?' The fact that fenoterol was not licensed in the United States might be taken as sufficient reason for its licence to be declined in New Zealand; yet the drug was licensed in most other Western countries and it would perhaps be harsh to blame the New Zealand Department of Health for making the same mistake.

It is reasonable to assume that fenoterol would never have been licensed in any country if the information that is now available had been available at the time of application for licensing – or even if the evidence which was available then had been properly and comprehensively considered. However, it can be very difficult to withdraw or restrict a drug once it has been licensed.

Given the combative environment in which debates on drug safety occur, it is important that the Ministry of Health should have the courage, and the resources, to act as an 'honest broker'. In particular, it should have advice from independent epidemiologists and pharmacologists. At the time of the fenoterol saga, the Department of Health lacked the technical expertise it needed to understand the issues. It was encouraged by the Minister of Health to seek independent advice from Professors Elwood and Skegg,[24] but for many months it did not have the courage to act on that advice. In some respects, the department was running true to form. In relation to the fenoterol controversy, Sandra Coney commented on the department's 'record of pussyfooting' and timidity in the face of pressure from drug companies.[25]

In response, the department could have argued that most 'scares' about the safety of drugs turn out to be false alarms, and it is dangerous to 'cry wolf' too often. No department of health will wish to place itself at risk of legal proceedings by taking action against a particular drug unless the evidence of hazard is very strong. Furthermore, the public is not well served if a department takes premature action on the basis of false alarms. However, the danger is that the frequency of such false alarms may create a sense of apathy, and that an appropriate response may not be made when real problems with drug safety do occur. When the appropriate response was to do nothing, the New Zealand Department of Health did this very well – it had plenty of experience. However, it was not so effective when real action was required. The Minister of Health, Helen Clark, later commented:

From the outset [the department] seemed strangely hesitant about accepting the implications of the research. My impression was that sensitivity to its relationship with the pharmaceutical company was being accorded a greater priority than interest in research which indicated that many of New Zealand's asthma deaths could be due to the use of fenoterol.[26]

THE MEDIA

Given the weakness of the Department of Health in dealing with the fenoterol controversy, the independent role of the media was crucial. However, despite the intense media interest in the issue, no reporters were asking what was behind the controversy, why the Department of Health's response was so weak, or why these findings had not previously been published by the Asthma Task Force, which had issued its first report on the epidemic back in 1985,[27] but had never reported which medicines had been used by the asthmatics who died. The reporters just printed the press statements they received, and didn't investigate further. They didn't even ask who had funded the meetings of 'independent experts' who criticised our studies.

There were many times when we took the criticism personally, or tried to 'sell' our case rather more than we should have done. It's important to avoid being seen as a campaigner. This is the 'kiss of death' to a scientific researcher, as you are then seen to be taking sides and picking the pieces of evidence that fit a particular point of view, rather than taking a balanced objective approach. I have sat in Ministry of Health meetings where a particular researcher has been described as an enthusiast – apparently it was intended as a criticism – and therefore as unbalanced and unreliable. There are some who argue that epidemiologists should play no part in public policy-making, because to do so compromises their objectivity and independence.

Yet, when your research is being criticised in the media, and you are the only person who knows the research in depth and knows that the criticisms are wrong, you feel obliged to respond. Just as the 'medical old boy network' had been very active in attempting to stifle and discredit our research, it was also important for us to contact the media to set the record straight and to encourage them to 'dig deeper'. So we did start talking regularly to reporters, returning their calls, and taking every opportunity to set the record straight. However, we had no chance against the Boehringer publicity machine, both because of our reluctance to become too heavily involved in the media, and because of our lack of time and resources to respond to the deluge of critical material and consensus panel reports.

Fortunately, things changed once Al Morrison got involved. Al was one of New Zealand's few investigative journalists, and worked for the *Dominion* in Wellington. We were in the same soccer team, which started in 1981 as a group of old friends who had been to Wellington's Victoria University together in the 1970s. Every Saturday afternoon the team would meet after the game in the club rooms in Vivian Street, in the former Chinese laundry where my wife had lived as a baby. After a few beers, Morrison would almost always get from one of the team a story that he could use the following week. One government department even hired a private investigator to try to find out where the leaks were coming from, but they never thought of checking out the soccer team. Unfortunately, Morrison was always strictly professional and the team members didn't always like his articles – as the season progressed fewer and fewer players would pass him the ball, particularly the two guys who worked for Treasury.

The team played for Wellington Diamond United. When we started with the club, Wynton Rufer played for the top team and was about to explode onto the international scene as part of New Zealand's 1982 World Cup campaign. Our team was a little less successful and didn't win a game for the first 18 months – the first game that I hadn't played in. My contribution to the team was appropriately recognised one year when I hurt my back, missed the second half of the season, and was given the award for 'most improved player'. However, the team had a fine collection of rogues and swindlers who could reinterpret the results of each game. Jamie Belich gave an historical analysis after every game to show that we had had a tactical victory even though we had had more goals scored against us than we had scored ourselves – he later used the same skills to reinterpret the New Zealand Wars. I kept the statistics and showed that we were constantly improving (we moved from an average 5–1 loss in the first season to an average 3–1 loss in the second). Paul Swain played striker and would lurk around the goal, telling us to get the ball and pass it to him so that he could score – skills that he later put to excellent use as a cabinet minister. Trace Hodgson was the team cartoonist and later achieved fame with his Jim Bolger 'potato head' cartoon. Peter Healey was perhaps the worst player ever to pull on a pair of soccer boots, but later did an excellent job as trade commissioner to India and then Vietnam.

One Saturday afternoon, just after the controversy blew up, Al

Morrison realised that I was one of the researchers involved in the controversy. After a few more beers, he got interested enough to investigate it further. For the first time there was a reporter who wanted to dig deep and find out the real story, rather than just reprinting press statements. Morrison started a series of pieces in the *Dominion* and *Sunday Times*, which were followed by other reporters, particularly Cate Brett at the *Christchurch Star*. Adele Broadbent was in my French evening class and also picked up the story with a series of items on National Radio.

RESEARCHERS

The problems with the media reporting of the fenoterol controversy illustrate the difficulties that hostile drug company publicity creates for researchers who have discovered evidence that a particular drug or chemical may be hazardous. In theory, any evidence of hazard should be made immediately available to the scientific community, and should have some influence on public health decision-making. In practice, researchers who have discovered evidence that a particular drug may be hazardous require very strong evidence, perseverance, a sense of humour and a good lawyer. Even then, there is the danger that, despite the best of intentions, researchers may over-react to the resulting wave of criticism and may tend to overstate the case against the drug, particularly if they consider that the drug company's criticisms are trivial, irrelevant or incorrect.

Other colleagues of mine have had their research attacked by drug companies and other vested interests. When asked, I always give them two pieces of advice. First, don't take the criticisms personally, even if the attacks are personal – if you get emotional and respond emotionally, you will always regret it the next day. Secondly, no matter how much you think your critics are bending the evidence, don't ever exaggerate the evidence that supports your case – keep to the facts and don't try to stretch them, try to be balanced, and be self-critical of your work, even if you know that your opponents will misuse this.

As Paul Stolley has written:

The pharmacoepidemiologist must develop a thick skin; this is not a field for timid souls. More important, the pharmacoepidemiologist must have some historical and ideologic anchorage and perspective to be able to understand the nature of the attacks. The history of drug regulation, the battle for effective drug efficacy and safety standards, and the connection of drug regulation with the sanitary movement of public health are all a part of the progressive tradition we have inherited as epidemiologists.[28]

Ultimately, the best approach for researchers is to address any criticisms in subsequent studies. Certainly, it was this approach which eventually led to the resolution of the fenoterol saga. It is important to remember that epidemiological research is usually a marathon rather than a sprint, and that to the extent that debates ever get settled, this happens over many years. When you are involved in a controversy like this it is important to ignore temporary problems and criticisms – real or imaginary, justified or unjustified – and to concentrate on the research. How is the controversy going to be viewed in 15 years time? How will you feel about it and about yourself? If education is what is left once everything we learned at school is forgotten, research is what is left when individual studies and publications are forgotten, or languish uncited in the archives. The stakes involved – people really did die during the epidemic in New Zealand, and more would have died if the epidemic had continued – make it even more important not to get side-tracked and to focus on doing more research and better studies. Most of the time, science wins in the end.

NOTES

INTRODUCTION

1. N. Pearce, 'Adverse reactions: the fenoterol saga'. In P. Davis (ed), *For health or profit: the pharmaceutical industry in New Zealand*. Auckland: Oxford University Press, 1992, pp.75–97; N. Pearce, 'Adverse reactions, social responses: a tale of two asthma mortality epidemics'. In P. Davis (ed), *Contested ground: public purpose and private interest in the regulation of prescription drugs*. New York: Oxford University Press, 1996, pp.57–75.

CHAPTER 1: SOME HISTORY

1. Y. Berra, *The Yogi book*. New York: Workman Press, 1998.
2. H.H. Salter, *On asthma: its pathology and treatment*. London, 1860.
3. N. Pearce, R. Beasley, C. Burgess and J. Crane, *Asthma epidemiology: principles and methods*. New York: Oxford University Press, 1998.
4. L. Unger and M.C. Harris, 'Stepping stones in allergy'. *Annals of Allergy* 32 (1974), pp.214–30.
5. L. Unger and M.C. Harris, p.217.
6. E.L. Keeney, 'The history of asthma from Hippocrates to Meltzer'. *Journal of Allergy & Clinical Immunology* 35 (1964), pp.15–226.
7. R. Ellul-Micallef, 'Asthma: a look at the past'. *British Journal of Diseases of the Chest* 70 (1976), pp.112–16.
8. M.R. Sears, H.H. Rea, G. De Boer, R. Beaglehole, et al., 'Accuracy of certification of deaths due to asthma – a national study'. *American Journal of Epidemiology* 124 (1986), pp.1004–11.
9. N. Pearce, J. Douwes and R. Beasley, 'Asthma'. In R. Detels, J. McEwen, R. Beaglehole, and H. Tanaka (eds), *Oxford textbook of public health*. Oxford: Oxford University Press, 2002, pp.1255–77.
10. K. Woodman, N. Pearce, R. Beasley, C. Burgess, et al., 'Albuterol and deaths from asthma in New Zealand from 1969 to 1976: a case-control study'. *Clinical Pharmacology & Therapeutics* 51 (1992), pp.566–71.
11. N. Pearce, R. Beasley, J. Crane and C. Burgess, 'Epidemiology of asthma mortality'. In S.T. Holgate and W.W. Busse (eds), *Asthma and rhinitis*. Oxford: Blackwell Scientific, 2001, pp.56–69.
12. M.J. Greenberg and A. Pines, 'Pressurized aerosols in asthma'. *British Medical Journal* 1 (1967), p.63.
13. R. Doll and A. Bradford Hill, 'Smoking and carcinoma of the lung: preliminary report'. *British Medical Journal* 2 (1950), pp.739–48; E.L. Wynder and E.A. Graham, 'Tobacco smoking as a possible etiologic factor in bronchiogenic carcinoma: a study of 684 proved cases'. *Journal of the American Medical Association* 143 (1950), pp.329–336.
14. R. Doll and A. Bradford Hill, 'Mortality in relation to smoking: ten years' observations of British doctors'. *British Medical Journal* i (1964), pp.1399–1410.
15. F.E. Speizer, R. Doll and P. Heaf, 'Observations on recent increase in mortality from asthma'. *British Medical Journal* 1968/1, pp.335–9.
16. H.H. Windom, C.D. Burgess, J. Crane, N. Pearce, et al., 'The self-administration of inhaled beta agonist drugs during severe asthma'. *New Zealand Medical Journal* 103 (1990), pp.205–7.

17. W.H.W. Inman and A.M. Adelstein, 'Rise and fall of asthma mortality in England and Wales in relation to use of pressurized aerosols'. *Lancet* 1969/2, pp.279–85.

18. F.E. Speizer, R. Doll, P. Heaf and L.B. Strang, 'Investigation into use of drugs preceding death from asthma'. *British Medical Journal* 1968/1, pp.339–43.

19. P.M. Fraser, F.E. Speizer, S.D.M. Waters, R. Doll, et al., 'The circumstances preceding death from asthma in young people in 1968 to 1969'. *British Journal of Diseases of the Chest* 65 (1971), pp.71–84.

20. N. Pearce, J. Crane, C. Burgess, R. Jackson, et al., 'Beta agonists and asthma mortality: deja vu'. *Clinical & Experimental Allergy* 21 (1991), pp.401–10.

21. J.M. Collins, D.G. McDevitt, R.G. Shanks and J.G. Swanton, 'The cardiotoxicity of isoprenaline during hypoxia'. *British Journal of Pharmacology* 36 (1969), pp.35–45.

22. R. Beasley, N. Pearce, J. Crane, H. Windom, et al., 'Asthma mortality and inhaled beta agonist therapy'. *Australian & New Zealand Journal of Medicine* 21 (1991), pp.753–63.

23. T.E. van Metre, 'Adverse effects of inhalation of excessive amounts of nebulized isoproterenol in status asthmaticus'. *Journal of Allergy* 43 (1969), pp.101–13.

24. P.D. Stolley, interview with Al Morrison, Wellington, 30 June 1989.

25. P.D. Stolley, interview.

26. J. Read, 'Pressurized aerosols in asthma'. *British Medical Journal* 1967/1, p.699.

27. R. Munro Ford, 'Asthma and aerosol sprays'. *Medical Journal of Australia* 1966/4, pp.667–8.

28. P.D. Stolley, interview.

29. P.D. Stolley, interview.

30. P.D. Stolley and R. Schinnar, 'Association between asthma mortality and isoproterenol aserosols: a review'. *Preventive Medicine* 7 (1978), pp.319–38.

31. P.D. Stolley, 'Why the United States was spared an epidemic of deaths due to asthma'. *American Review of Respiratory Disease* 105 (1972), pp.883–90.

32. P.D. Stolley, interview.

33. P.D. Stolley, 'Asthma deaths and bronchodilator aerosols – reply'. *American Review of Respiratory Disease* 107 (1973), pp.1078–79.

34. B. Gandevia, 'Pressurized sympathomimetic aerosols and their lack of relationship to asthma mortality in Australia'. *Medical Journal of Australia* i (1973), pp.273–7.

35. A.H. Campbell, 'Mortality from asthma and bronchodilator aerosols'. *Medical Journal of Australia* 1976/1, pp.386–91.

36. P.D. Stolley, 'Why the United States was spared an epidemic of deaths', p.889.

37. P.D. Stolley, interview.

38. G.R. Venning, 'Identification of adverse reactions to new drugs. I. What have been the important adverse reactions since thalidomide?' *British Medical Journal* 286 (1983), pp.199–202.

39. 'Asthma deaths: a question answered'. *British Medical Journal* 4 (1972), pp.443–4.

CHAPTER 2: HISTORY IS REWRITTEN

1. P.D. Stolley, interview with Al Morrison, Wellington, 30 June 1989.

2. P.D. Stolley and R. Schinnar, 'Association between asthma mortality and isoproterenol aserosols: a review'. *Preventive Medicine* 7 (1978), pp.319–38.

3. J. Read, 'The reported increase in mortality from asthma: a clinico-functional analysis'. *Medical Journal of Australia* 1968/1, pp.879–91.

4. 'Fatal asthma'. *Lancet* 1979/2, pp.337–8.

5. L.G. Olson, 'Acute severe asthma: what to do until the ambulance arrives'. *New Ethicals* 25 (1988), pp.105–16.

6. P.M. Fraser, F.E. Speizer, S.D.M. Waters, R. Doll, et al., 'The circumstances preceding death from asthma in young people in 1968 to 1969'. *British Journal of Diseases of the Chest* 65 (1971), pp.71–84.

7. P.R. Patrick and J.I. Tonge, 'Asthma and aerosol sprays'. *Medical Journal of Australia*. 1966/2, p.668.

8. N. Pearce, J. Crane, C. Burgess, R. Jackson, et al., 'Beta agonists and asthma

mortality: déjà vu'. *Clinical & Experimental Allergy* 21 (1991), pp.401–10.

9. L. Hendeles and M. Weinberger, 'Nonprescription sale of inhaled metaproterenol – deja vu'. *New England Journal of Medicine* 310 (1984), pp.207–8.

10. W.H.W. Inman and A.M. Adelstein, 'Rise and fall of asthma mortality in England and Wales in relation to use of pressurized aerosols'. *Lancet* 1969/2, pp.279–285.

11. W.M. Wardell, 'Nonprescription sales of metaproterenol aerosols'. *New England Journal of Medicine* 311 (1984), pp.405–6.

12. M. Weinberger and L. Hendeles, 'Nonprescription sales of metaproterenol aerosols'. *New England Journal of Medicine* 311 (1984), p.406.

13. ERI, and Walker in particular, had played a key role when Johns-Manville, the major United States asbestos company, filed for a 'Chapter 11' bankruptcy, which enabled it to avoid liability for asbestos-related disease but still continue operating. (P. Brodeur, *Outrageous misconduct: the asbestos industry on trial.* New York: Pantheon, 1985). ERI had been hired by the industry to produce estimates of the future number of cases of asbestos disease and the number of potential claims and lawsuits. However, Walker's initial estimates were 'too high' and would have required the company to file for a 'Chapter 7' bankruptcy, thus closing down the company. Walker was therefore asked to revise his estimates. In subsequent bankruptcy proceedings, Walker conceded that 'I was asked [that] . . . whenever I had to choose between two equally plausible assumptions, I should choose the assumption which led to a smaller number of cases of disease.' Walker's final estimate was 139,000 cases of asbestos disease during 1980–2009, considerably lower than his original estimate of 230,000 cases. This was then combined with estimates of the proportion of cases which might result in lawsuits, and with estimates of the average cost per lawsuit, to obtain estimates of the likely cost of future litigation. This was then used to justify the Chapter 11 bankruptcy, which was filed in August 1982. As a result, the company gained an automatic stay in 11,000 asbestos disease lawsuits which were pending, and tens of thousands of asbestos workers who would have sought compensation in the future were effectively denied the opportunity to do so.

14. S.F. Lanes and A.M. Walker, 'Do pressurized bronchodilator aerosols cause death among asthmatics?' *American Journal of Epidemiology* 125 (1987), pp.755–60.

15. F.E. Speizer, R. Doll and P. Heaf, 'Observations on recent increase in mortality from asthma'. *British Medical Journal* 1968/1, pp.335–9.

16. S. Greenland, 'Science versus advocacy: the challenge of Dr Feinstein'. *Epidemiology* 1 (1990), pp.64–72; D.A. Savitz, S. Greenland, P.D. Stolley and J.L. Kelsey, 'Scientific standards of criticism: a reaction to "Scientific standards in epidemiologic studies of the menace of daily life." by A.R. Feinstein'. *Epidemiology* 1 (1990), pp.78–83.

17. J.M. Esdaile, A.R. Feinstein and R.I. Horwitz, 'A reappraisal of the United Kingdom epidemic of fatal asthma'. *Archives of Internal Medicine* 147 (1987), pp.543–9.

18. P.D. Stolley, interview.

19. P.D. Stolley, interview.

20. P.D. Stolley, interview.

21. M.A. Jenkins, S.F. Hurley, G. Bowes and J.J. McNeil, 'Use of antiasthmatic drugs in Australia'. *Medical Journal of Australia.* 153 (1990), pp.32–38.

22. S.R. Benatar, 'Fatal asthma'. *New England Journal of Medicine* 314 (1986), pp.423–9.

23. P.D. Stolley and T. Lasky, 'The bellman always rings thrice'. *Annals of Internal Medicine* 118 (1993), p.158.

24. P.D. Stolley, interview.

CHAPTER 3: HISTORY REPEATS ITSELF

1. G. Santayana, *Life of reason: reason in common sense.* Vol. 1. New York: Scribner's, 1905.

2. J.D. Wilson, D.C. Sutherland and A.C. Thomas, 'Has the change to beta-agonists

combined with oral theophylline increased cases of fatal asthma?' *Lancet* 1981/1, pp.1235–7.

3. J.D. Wilson, Midweek, TVNZ documentary, Auckland, 1981.
4. A. Flatt, C. Burgess, H. Windom, R. Beasley, et al., 'The cardiovascular effects of inhaled fenoterol alone and during treatment with oral theophylline'. *Chest* 96 (1989), pp.1317–20.
5. J.D. Wilson, Midweek, TVNZ documentary.
6. N. Pearce, 'Adverse reactions, social responses: a tale of two asthma mortality epidemics'. In P. Davis (ed), *Contested ground: public purpose and private interest in the regulation of prescription drugs*. New York: Oxford University Press, 1996, pp.57–75.
7. W.H.W. Inman and A.M. Adelstein, 'Rise and fall of asthma mortality in England and Wales in relation to use of pressurized aerosols'. *Lancet* 1969/2, pp.279–85.
8. R. Jackson, interview by Elisabeth Heseltine, Auckland, 23 February 1991.
9. R.T. Jackson, R. Beaglehole, H.H. Rea and D.C. Sutherland, 'Mortality from asthma: a new epidemic in New Zealand'. *British Medical Journal* 285 (1982), pp.771–4.
10. 'Death from asthma in New Zealand'. *New Zealand Medical Journal* 95 (1982), pp.692–3.
11. I.W.B. Grant, 'Asthma in New Zealand'. *British Medical Journal* 286 (1983), pp.374–7.
12. I.W.B. Grant to Asthma Task Force, 11 January 1985. Copy in author's possession.
13. D.C. Sutherland and H.H. Rea, 'Asthma in New Zealand'. *British Medical Journal* 286 (1983), pp.717–18.
14. I.W.B. Grant, 'Asthma in New Zealand – Reply'. *British Medical Journal* 286 (1983), p.718.
15. P.E. Thornley and K.P. Dawson, 'Asthma in New Zealand'. *British Medical Journal* 286 (1983), pp.890–1.
16. T.V. O'Donnell, 'Asthma in New Zealand: a response'. *New Zealand Medical Journal* 96 (1983), pp.163–5.
17. G. Keating, E.A. Mitchell, R. Jackson, R. Beaglehole, et al., 'Trends in sales of drugs for asthma in New Zealand, Australia and the United Kingdom, 1975–81'. *British Medical Journal* 289 (1984), pp.348–51.
18. P.D. Stolley, 'Why the United States was spared an epidemic of deaths due to asthma'. *American Review of Respiratory Disease* 105 (1972), pp.883–90.
19. P.D. Stolley, interview with Al Morrison, Wellington, 30 June 1989.
20. H.H. Rea, R. Scragg, R. Jackson, R. Beaglehole, et al., 'A case-control study of deaths from asthma'. *Thorax* 41 (1986), pp.833–9.
21. M.R. Sears, H.H. Rea, R. Beaglehole, A.J.D. Gillies, et al., 'Asthma mortality in New Zealand: a two year national study'. *New Zealand Medical Journal* 98 (1985), pp.271–5.
22. M.R. Sears and R. Beaglehole, 'Asthma morbidity and mortality: New Zealand'. *Journal of Allergy & Clinical Immunology* 80 (1987), pp.383–8.
23. R. Beaglehole, R. Jackson, M. Sears and H.H. Rea, 'Asthma mortality in New Zealand: a review with some policy implications'. *New Zealand Medical Journal* 100 (1987), pp.231–4.
24. R.J. Mackay and J.H. van der Westhuyzen, 'Asthma in New Zealand: some answers, more questions'. *New Zealand Medical Journal* 101 (1988), pp.835–6.
25. M.R. Sears, 'Asthma in New Zealand: some answers, more questions'. *New Zealand Medical Journal* 101 (1988), pp.598–600.

CHAPTER 4: 'MAYBE THESE DRUGS ARE NOT ALL THE SAME'

1. A. Morrison, 'Controversy rages over asthma drug'. *Sunday Times*, Wellington, 2 July 1989.
2. M.R. Sears, H.H. Rea, R. Beaglehole, A.J.D. Gillies, et al., 'Asthma mortality in New Zealand: a two year national study'. *New Zealand Medical Journal* 98 (1985), pp.271–5.

3. N. Pearce, J. Crane, C. Burgess, R. Jackson, et al., 'Beta agonists and asthma mortality: déjà vu'. *Clinical & Experimental Allergy* 21 (1991), pp.401–10.
4. G. Keating, E.A. Mitchell, R. Jackson, R. Beaglehole, et al., 'Trends in sales of drugs for asthma in New Zealand, Australia and the United Kingdom, 1975–81'. *British Medical Journal* 289 (1984), pp.348–51.
5. Morrison.
6. J.D. Wilson, D.C. Sutherland and A.C. Thomas, 'Has the change to beta-agonists combined with oral theophylline increased cases of fatal asthma?' *Lancet* 1981/1, pp.1235–7.
7. D. Dawson, 'Physicians back asthma findings'. *Sunday Times*, Wellington, 30 April 1989.
8. R. Beasley, J. Crane, C. Burgess, N.E. Pearce, et al., 'Fenoterol and severe asthma mortality'. *New Zealand Medical Journal* 102 (1989), pp.294–5.
9. R. Beasley, C. Burgess, N. Pearce, K. Woodman, et al., 'Confounding by severity does not explain the association between fenoterol and asthma death'. *Clinical & Experimental Allergy* 24 (1994), pp.660–8.
10. J. Crane, interview with Elisabeth Heseltine, Wellington, 20 February 1991.
11. I.W.B. Grant, 'Fenoterol and asthma deaths in New Zealand'. *New Zealand Medical Journal* 103 (1990), pp.160–1.
12. C. Burgess, interview with Elisabeth Heseltine, Wellington, 20 February 1991.
13. C. Burgess, interview.
14. J. Crane, C. Burgess and R. Beasley, 'Cardiovascular and hypokalaemic effects of inhaled salbutamol, fenoterol and isoprenaline'. *Thorax* 44 (1989), pp.136–40.
15. C. Burgess, interview.
16. M.K. Tandon, 'Cardiopulmonary effects of fenoterol and salbutamol aerosols'. *Chest* 77 (1980), pp.429–31.
17. R. Beasley, N. Pearce, J. Crane, H. Windom, et al., 'Asthma mortality and inhaled beta agonist therapy'. *Australian & New Zealand Journal of Medicine* 21 (1991), pp.753–63.
18. P. Bremner, R. Siebers, J. Crane, R. Beasley, et al., 'Partial vs full b–receptor agonism, a clinical study of inhaled albuterol and fenoterol'. *Chest* 109 (1996), pp.957–62.
19. M.R. Sears, H.H. Rea, H. Fenwick, A.J.D. Gillies, et al., '75 deaths in asthmatics prescribed home nebulisers'. *British Medical Journal* 294 (1987), pp.477–80.
20. M.R. Sears, H.H. Rea, H. Fenwick, A.J.D. Gillies, et al., p.479.
21. R. Beasley, interview with Elisabeth Heseltine, Wellington, 20 February 1991.
22. R. Beasley, interview.

CHAPTER 5: THE FIRST STUDY

1. R. Beasley to T.V. O'Donnell, 14 December 1987. Copy in author's possession.
2. R. Beasley to T.V. O'Donnell, 22 December 1987. Copy in author's possession.
3. R. Beasley, interview with Elisabeth Heseltine, Wellington, 20 February 1991.
4. M.R. Sears, H.H. Rea, R. Beaglehole, A.J.D. Gillies, et al., 'Asthma mortality in New Zealand: a two year national study'. *New Zealand Medical Journal* 98 (1985), pp.271–5.
5. R. Beasley, interview.
6. J. Crane, N. Pearce, A. Flatt, C. Burgess, et al., 'Prescribed fenoterol and death from asthma in New Zealand, 1981–83: case-control study'. *Lancet* 1989/1, pp.917–22.
7. M.R. Sears, H.H. Rea, H. Fenwick, A.J.D. Gillies, et al., '75 deaths in asthmatics prescribed home nebulisers'. *British Medical Journal* 294 (1987), pp.477–80.
8. R. Beasley to T.V. O'Donnell, 8 April 1988. Copy in author's possession.
9. R. Beasley to T.V. O'Donnell, 13 April 1988. Copy in author's possession.
10. R. Beasley to H. Rea, 13 April 1988. Copy in author's possession.
11. J. Crane, interview with Elisabeth Heseltine, Wellington, 20 February 1991.
12. J. Crane, interview.

13. C. Burgess, interview with Elisabeth Heseltine, Wellington, 20 February 1991.
14. N. Pearce, R. Beasley, C. Burgess and J. Crane, *Asthma epidemiology: principles and methods*. New York: Oxford University Press, 1998.
15. R.C. Strunk, D.A. Mrazek, G.S. Wolfson Fuhrmann and J.F. LaBrecque, 'Physiologic and psychologic characteristics associated with deaths due to asthma in childhood'. *Journal of the American Medical Association* 254 (1985), pp.1193–8.
16. H.H. Rea, R. Scragg, R. Jackson, R. Beaglehole, et al., 'A case-control study of deaths from asthma'. *Thorax* 41 (1986), pp.833–9.
17. M.R. Sears, H.H. Rea, G. De Boer, R. Beaglehole, et al., 'Accuracy of certification of deaths due to asthma – a national study'. *American Journal of Epidemiology* 124 (1986), pp.1004–11.
18. N. Pearce and M.J. Hensley, 'Epidemiologic studies of beta agonists and asthma deaths'. *Epidemiologic Reviews* 20 (1998), pp.173–86.
19. H.H. Rea, R. Scragg, R. Jackson, R. Beaglehole, et al., p.837.
20. J.D. Wilson, memorandum attached to letter to R. Beasley, 13 October 1988. Copy in author's possession.
21. J. Crane, C. Burgess and R. Beasley, 'Cardiovascular and hypokalaemic effects of inhaled salbutamol, fenoterol and isoprenaline'. *Thorax* 44 (1989), pp.136–40.
22. Anonymous review of grant application to the Asthma Foundation of New Zealand, 1988. Copy in author's possession.
23. D. Wilson to R. Beasley, 20 July 1988. Copy in author's possession.

CHAPTER 6: GETTING THE PAPER PUBLISHED

1. J.D. Watson, *The double helix: a personal account of the discovery of the structure of DNA*. New York: Simon & Schuster, 1968.
2. M.R. Sears to R. Beasley, 5 December 1989. Copy in author's possession.
3. H.H. Rea to R. Beasley, 8 December 1988. Copy in author's possession.
4. J. Hodge to R. Beasley, 7 December 1988. Copy in author's possession.
5. D. Wilson to R. Beasley, 5 December 1988. Copy in author's possession.
6. L.A. Malcolm to T.V. O'Donnell, 9 December 1988. Copy in author's possession.
7. S.J. Leeder, unpublished minutes of a meeting of asthma investigators to discuss progress of a study of asthma deaths in New Zealand and their association with the use of inhaled fenoterol, Wellington, 21 December 1988. Copy in author's possession.
8. J. Crane, interview with Elisabeth Heseltine, Wellington, 20 February 1991.
9. D. Jones to R. Beasley, 25 January 1989. Copy in author's possession.
10. H.H. Rea, review of Auckland and Northland controls, 27 January 1989. Copy in author's possession.
11. R. Beasley to T.V. O'Donnell, 27 January 1989. Copy in author's possession.
12. 'Questions continue on sale of asthma drug'. *Christchurch Star*, Christchurch, 9 December 1989; D.G. Salmond, internal Department of Health memorandum, 6 January 1989. Ministry of Health archives.
13. G. Salmond to J.D. Wilson, 7 February 1989. Copy in author's possession.
14. G. Salmond to editor of the *Lancet*, 7 February 1989. Copy in author's possession.
15. R.T. Jackson, R. Beaglehole, H.H. Rea and D.C. Sutherland, 'Mortality from asthma: a new epidemic in New Zealand'. *British Medical Journal* 285 (1982), pp.771–4.
16. 'Fenoterol hypothesis: study weaknesses'. Auckland: Boehringer Ingelheim, 1989. Copy in author's possession.
17. J. Haas memorandum to V. Hartmann, 8 February 1989. Copy in author's possession.
18. D. Sharp to J. Crane, 20 February 1989. Copy in author's possession.
19. J.M. Esdaile, A.R. Feinstein and R.I. Horwitz, 'A reappraisal of the United Kingdom epidemic of fatal asthma'. *Archives of Internal Medicine* 147 (1987), pp.543–9; S.F. Lanes and A.M. Walker, 'Do pressurized bronchodilator aerosols cause death among asthmatics?' *American Journal of Epidemiology* 125 (1987), pp.755–60.

20. A.R. Feinstein to J.D. Wilson, 16 February 1989. Copy in author's possession.
21. S.F. Lanes and A.M. Walker to J. Wahl, 15 February 1989. Copy in author's possession.
22. A.S. Buist to J. Wahl, 16 February 1989. Copy in author's possession.
23. A.S. Rebuck to J.D. Wilson, March 1989. Copy in author's possession.
24. W.O. Spitzer and R.I. Horwitz, Preliminary review of 'Prescribed fenoterol anddeath from asthma in New Zealand 1981–1983: A case-control study (J. Crane et al.)', 1 March 1989. Copy in author's possession.
25. J.B.L. Howell and W.E. Waters to D.M. Humphreys, Boehringer Ingelheim, 17 March 1989. Copy in author's possession.
26. P. Burney report for Boehringer Ingelheim on 'Prescribed fenoterol and death from asthma in New Zealand, 1981–1983: a case-control study', 27 February 1989. Copy in author's possession.
27. D.L. Sackett, G. Browman and L. Marrett, manuscript review, 3 March 1989. Copy in author's posession.
28. G. Browman to V. Hartmann, 17 March 1989. Copy in author's possession.
29. F.E. Speizer, R. Doll and P. Heaf, 'Observations on recent increase in mortality from asthma'. *British Medical Journal* 1968/1, pp.335–9; F.E. Speizer, R. Doll, P. Heaf and L.B. Strang, 'Investigation into use of drugs preceding death from asthma'. *British Medical Journal* 1968/1, pp.339–43.
30. 'Asthma medication and mortality: a submission to the Adverse Drug Reactions Committee, 17 February 1989'. Auckland: Boehringer Ingelheim, 1989; 'Submission from Boehringer Ingelheim Ltd to Medicines Adverse Reactions Committee, 13 March 1989'. Auckland: Boehringer Ingelheim, 1989.
31. R. Beasley, J. Crane, C. Burgess, N.E. Pearce, et al., 'Fenoterol and severe asthma mortality'. *New Zealand Medical Journal* 102 (1989), pp.294–5.
32. T.V. O'Donnell, H.H. Rea, P.E. Holst and M.R. Sears, submission from Asthma Task Force to MARC, 21 February 1989. Copy in author's possession.
33. M. Lebowitz review of 'Prescribed fenoterol and death from asthma in New Zealand, 1981–83: a case-control study'. Copy in author's possession.
34. M. Sears to T.V. O'Donnell, 8 February 1989. Copy in author's possession.
35. R. Beasley, interview with Elisabeth Heseltine, Wellington, 20 February 1991.
36. D. Sharp to J. Crane, 3 March 1989. Copy in author's possession.
37. W. Thompson to Ichiro Kawachi, 30 June 1989. Copy in author's possession.
38. R. Beasley, C. Burgess, J. Crane and N. Pearce to J. D. Wilson, 4 April 1989.
39. J. Crane, C. Burgess and R. Beasley, 'Cardiovascular and hypokalaemic effects of inhaled salbutamol, fenoterol and isoprenaline'. *Thorax* 44 (1989), pp.136–40.
40. D.L. Sackett, H.S. Shannon and G.W. Browman, 'Fenoterol and Fatal Asthma'. *Lancet* 335 (1990), pp.45–46.
41. A.S. Buist, P.G.J. Burney, A.R. Feinstein, R.I. Horwitz, et al., 'Fenoterol and fatal asthma'. *Lancet* 1989/1, p.1071; S. Buist, P. Burney, R. Horwitz, S. Lanes, et al., 'Consensus report: an appraisal of a manuscript by J. Crane et al'. New York, 1989.
42. H.H. Windom, C.D. Burgess, J. Crane, N. Pearce, et al., 'The self-administration of inhaled beta agonist drugs during severe asthma'. *New Zealand Medical Journal* 103 (1990), pp.205–7.
43. K.T. Copeland, H. Checkoway, A.J. McMichael and R.H. Holbrook, 'Bias due to misclassification in the estimation of relative risk'. *American Journal of Epidemiology* 105 (1977), pp.488–95.
44. R. Beasley, C. Burgess, N. Pearce, K. Woodman, et al., 'Confounding by severity does not explain the association between fenoterol and asthma death'. *Clinical & Experimental Allergy.* 24 (1994), pp.660–8.
45. J. Crane, N. Pearce, A. Flatt, C. Burgess, et al., 'Prescribed fenoterol and death from asthma in New Zealand, 1981–83: case-control study'. *Lancet* 1989/1, p.920.
46. J. Crane, N. Pearce, A. Flatt, C. Burgess, et al., 'Prescribed fenoterol and death from asthma in New Zealand, 1981–83: case-control study'. *Lancet* 1989/1, p.918.
47. R. Doll to R. Jackson, 19 December 1988. Copy in author's possession.

48. A.H. Smith to G. Reeves, 16 March 1989. Copy in author's possession.
49. R. Beasley, C. Burgess, J. Crane, R. Jackson, et al. to D. Sharp, 17 March 1989. Copy in author's possession.
50. J. Crane, N. Pearce, C. Burgess and C.R.W. Beasley, Unpublished letter to editor of the *Lancet*, March 1989. Copy in author's possession.
51. D. Sharp, reply to J. Crane et al., March 1989. Copy in author's possession.

CHAPTER 7: ADVERSE REACTIONS
1. D.G. Salmond, internal Department of Health memorandum, 15 March 1989. Ministry of Health archives.
2. H. Clark, 'Pharmaceutical costs and regulation: from the minister's desk'. In P.B. Davis (ed), *For health or profit: the pharmaceutical industry in New Zealand*. Auckland: Oxford University Press, 1992.
3. C. Cowie, 'Quick asthma study back-up unlikely'. *Evening Post*, Wellington, 1 May 1989.
4. R.D. Scobie and J.D. Wilson to Director-General of Health, 27 April 1989. Ministry of Health archives.
5. C. Hauke and V. Hartmann to Director-General of Health, 3 May 1989. Copy in author's possession.
6. I.R. Edwards to K.H. Goh, Department of Health, 20 February 1989. Copy in author's possession.
7. J.M. Elwood and D.C.G. Skegg, 'Review of studies relating to prescribed fenoterol and death from asthma in New Zealand'. Wellington: Department of Health, 1989.
8. S. Leeder to R. Beasley, 8 March 1989. Copy in author's possession.
9. J. Crane, N. Pearce, A. Flatt, C. Burgess, et al., 'Prescribed fenoterol and death from asthma in New Zealand, 1981–83: case-control study'. *Lancet* 1989/1, pp.917–22.
10. H. Clark, p.71.
11. 'Manufacturer rejects findings as faulty'. *Dominion*, Wellington, 29 April 1989.
12. S. Buist, P. Burney, R. Horwitz, S. Lanes, et al., 'Consensus report: an appraisal of a manuscript by J. Crane et al'. New York, 1989.
13. C. Cowie, 'Asthma drug maker under pressure'. *Evening Post*, Wellington, 2 May 1989.
14. 'Serious flaws in study, says panel'. *Dominion*, Wellington, 29 April 1989.
15. 'Asthma study flawed, say foreign experts'. *Auckland Star*, Auckland, 19 May 1989.
16. 'Asthma drug warning sent to doctors'. *Dominion*, Wellington, 1 May 1989.
17. 'Panel says fenoterol study "seriously flawed"'. *Australian Doctor*, 16 May 1989.
18. J.R. Wickens to J.D. Wilson, 13 May 1989. Copy in author's possession.
19. I. Wilkins, 'Asthmatics in dark as drug study widens research row'. *Sunday Star*, Auckland, 7 May 1989.
20. R. Drent, 'Harm to doctors' training feared'. *Dominion*, Wellington, 12 July 1989.
21. A. Morrison, 'Controversy rages over asthma drug'. *Sunday Times*, Wellington, 2 July 1989.
22. 'Questions continue on sale of asthma drug'. *Christchurch Star*, Christchurch, 9 December 1989.
23. *Christchurch Star*, 9 December 1989.
24. H. Clark, p.72.
25. J.M. Elwood and D.C.G. Skegg, Press statement, May 1989. Copy in author's possession.
26. A.S. Buist, P.G.J. Burney, A.R. Feinstein, R.I. Horwitz, et al., 'Fenoterol and fatal asthma'. *Lancet* 1989/1, p.1071; T.V. O'Donnell, H.H. Rea, P.E. Holst and M.R. Sears, 'Fenoterol and fatal asthma'. *Lancet* 1 (1989), pp.1070–1; N. Pearce, J. Crane, C. Burgess, R. Beasley, et al., 'Fenoterol and Asthma Mortality'. *Lancet* 1 (1989), pp.1196–7.
27. 'Delay may put lives at risk'. *Evening Post*, Wellington, 3 May.
28. *Sunday Star*, 7 May 1989.

29. 'Company's move "pre-empts" report'. *Dominion*, Wellington, 1 May 1989.
30. C. Cowie, 'Public health group seeks asthma inquiry'. *Evening Post*, Wellington, 8 May 1989.
31. 'Dispute leaves asthmatics confused'. *Auckland Star*, Auckland, 7 May 1989.
32. 'Asthmatics please read this'. *Evening Post*, Wellington, 2 May 1989.
33. I. Horswell, 'Ape deaths link with asthma drug'. *Sunday Star*, Auckland, 4 May 1989.
34. E. Cohen and A. Van As, 'A haemodynamic study on fenoterol (Berotec) in the baboon'. *Medical Proceedings* (1972), pp.24–27.
35. D. Dawson, 'Physicians back asthma findings'. *Sunday Times*, Wellington, 30 April 1989.
36. H. Kundig, 'Preliminary pharmacological and toxicological studies in the baboon (Papio ursinus) on a new β–2 adrenergic stimulant, fenoterol (Berotec)'. *Medical Proceedings* (1972), pp.9–14.
37. D. Chisholm, 'Asthma drug given to delay labour'. *Sunday Star*, Auckland, 7 May.
38. J. Crane, N. Pearce, C. Burgess and R. Beasley, 'Fenoterol and fear of flying'. *New Zealand Medical Journal* 102 (1989), pp.514.
39. *Sunday Star*, 7 May 1989.
40. R. Beasley, C. Burgess, J. Crane, R. Jackson, et al., submission to MRC meeting on fenoterol and asthma mortality. Wellington: Wellington School of Medicine, 1989.
41. R. Drent, 'Asthma study unsure – council'. *Dominion*, Wellington, 26 May 1989.

CHAPTER 8: THE SECOND STUDY

1. N. Pearce, J. Grainger, M. Atkinson, J. Crane, et al., 'Case-control study of prescribed fenoterol and death from asthma in New Zealand, 1977–81'. *Thorax* 45 (1990), pp.170–5.
2. Unpublished transcript of discussion in Workshop on Pharmacology and Clinical Epidemiology, University of Newcastle, 1989, unpaginated. Copy in author's possession.
3. G. Ray, 'Researchers say asthma puffer drugs need rethink'. *Newcastle Herald*, Newcastle, Australia, 29 June 1989.
4. R. Drent, 'Asthma drug blamed again'. *Dominion*, Wellington, 29 June 1989.
5. A. Morrison, 'Controversy rages over asthma drug'. *Sunday Times*, Wellington, 2 July 1989.

CHAPTER 9: MIXED REACTIONS

1. J. McEwen to J. Crane, 5 July 1989. Copy in author's possession.
2. D.W. Girvan to N.E. Pearce, 6 July 1989. Copy in author's possession.
3. Draft letter by Macalister Mazengarb to D.W. Girvan, 7 July 1989. Copy in author's possession.
4. D.H. Lawson to Ms Kathlyn J. Ronaldson, Department of Health, 18 July 1989. Copy in author's possession.
5. 'Questions continue on sale of asthma drug'. *Christchurch Star*, Christchurch, 9 December 1989.
6. 'Fenoterol investigation uncovers pharmaceutical company demands'. *PHA News*, August 1989.
7. I. Kawachi to Minister of Health, 25 October 1989. Copy in author's possession.
8. 'Edwards turns down funds'. *Christchurch Star*, Christchurch, 9 December 1989.
9. H. Clark, 'Pharmaceutical costs and regulation: from the minister's desk'. In P.B. Davis (ed), *For health or profit: the pharmaceutical industry in New Zealand*. Auckland: Oxford University Press, 1992.
10. A. Morrison, 'Power plays in the fight to save a suspect drug'. *Dominion*, Wellington, 22 September 1989.
11. D. Clark to Minister of Health, 20 July 1989. Copy in author's possession.
12. N. Pearce to W. Thompson, Ministry of Health, 17 July 1989. Copy in author's possession.
13. A.J. Woolcock. 'Aerosol bronchodilator therapy in asthma'. In *10th Asia Pacific*

Congress on Diseases of the Chest. Amsterdam: Excerpta Medica, Elsevier, 1985.
14. A.J. Woolcock, M.P. Alpers and V.M. Hurry, 'An epidemic of asthma in Papua-New Guinea'. *Australian and New Zealand Journal of Medicine* 18 (1988), p.544.
15. J. Crane, R. Beasley, N. Pearce and C. Burgess to P. Burney, 25 July 1989. Copy in author's possession.
16. J.V. Hodge, 'Fenoterol and asthma mortality: report on attendance at a meeting in Los Angeles on 29 July 1989'. Auckland: Medical Research Council, 1989.
17. S. Buist, P. Burney, P. Ernst, R. Horwitz, et al., 'Consensus report: An appraisal of a manuscript by N. Pearce et al'. Los Angeles, CA: Boehringer Ingelheim, 1989.
18. J.A. Borrows, Requests for applications in asthma research to be funded by Boehringer Ingelheim. Auckland: Medical Research Council, 1989.
19. R. Beasley to J.V. Hodge, 14 September 1989. Copy in author's possession.
20. J.V. Hodge, 'Drug-related asthma mortality: notes on response to call for comment on Boehringer RFPs'. Auckland: Medical Research Council, 1989.
21. D. Bandaranayake to J. Borrows, 5 October 1989. Copy in author's possession.
22. A. Morrison, 'Council quits research funded by drug firm'. *Dominion*, Wellington, 17 October 1989.
23. N. Pearce to R.G. Robinson, 12 December 1989. Copy in author's possession.
24. C. Poole, S.F. Lanes and A.M. Walker, 'Fenoterol and fatal asthma'. *Lancet* 335 (1990), p.920.
25. N. Pearce, J. Grainger, M. Atkinson, J. Crane, et al., 'Case-control study of prescribed fenoterol and death from asthma in New Zealand, 1977–81'. *Thorax* 45 (1990), pp.170–5.
26. J. Last, 'Epidemiology and ethics'. *Lancet* 336 (1990), p.497.
27. 'Questions continue on sale of asthma drug'. *Christchurch Star*, 9 December 1989.
28. J.M. Elwood, 'Prescribed fenoterol and deaths from asthma in New Zealand – second report'. Dunedin: University of Otago, 1989.
29. 'Asthma drug use to be restricted'. *Evening Post*, Wellington, 21 December 1989.
30. P.D. Stolley, fax to Pearce, Beasley et al., 21 December 1989. Copy in author's possession.

CHAPTER 10: POSITIVE REACTIONS

1. 'Fenoterol on the way out'. *Dominion*, Wellington, 27 December 1989.
2. S. Quayle, 'Warning on asthma drug'. *Australian Doctor*, 16 March 1989.
3. *Australian Doctor*, 16 March 1989.
4. J. Grainger, K. Woodman, N. Pearce, J. Crane, et al., 'Prescribed fenoterol and death from asthma in New Zealand, 1981–7: a further case-control study'. *Thorax* 46 (1991), pp.105–11.
5. R. Beasley, N. Pearce, J. Crane, H. Windom, et al., 'Asthma mortality and inhaled beta agonist therapy'. *Australian & New Zealand Journal of Medicine* 21 (1991), pp.753–63.
6. G. Keating, E.A. Mitchell, R. Jackson, R. Beaglehole, et al., 'Trends in sales of drugs for asthma in New Zealand, Australia and the United Kingdom, 1975–81'. *British Medical Journal* 289 (1984), pp.348–51.
7. N. Pearce, R. Beasley, J. Crane and C. Burgess, 'Epidemiology of asthma mortality'. In S.T. Holgate and W.W. Busse (eds), *Asthma and rhinitis*. Oxford: Blackwell Scientific, 2001, pp.56–69.
8. M.R. Sears and D.R. Taylor, 'The beta–2 agonist controversy: observations, explanations and relationship to asthma epidemiology'. *Drug Safety* 11 (1994), pp.259–83.
9. M.R. Sears, D.R. Taylor, C.G. Print, D.C. Lake, et al., 'Regular inhaled beta-agonist treatment in bronchial asthma'. *Lancet* 336 (1990), pp.1391–6.
10. C.S. Wong, I.D. Pavord, J. Williams, J.R. Britton, et al., 'Bronchodilator, cardiovascular, and hypokalemic effects of fenoterol, salbutamol, and terbutaline in asthma'. *Lancet* 336 (1990), pp.1396–9.
11. 'Beta–2 agonists in asthma: relief, prevention, morbidity'. *Lancet* 336 (1990),

pp.1391–6.

12. W.O. Spitzer, S. Suissa, P. Ernst, R.I. Horwitz, et al., 'The use of beta-agonists and the risk of death and near death from asthma'. *New England Journal of Medicine* 326 (1992), pp.501–6; S. Suissa, P. Ernst, J.F. Boivin, R.I. Horwitz, et al., 'A cohort analysis of excess mortality in asthma and the use of inhaled beta-agonists'. *American Journal of Respiratory and Critical Care Medicine* 149 (1994), pp.604–10.

13. K. Woodman, N. Pearce, R. Beasley, C. Burgess, et al., 'Albuterol and deaths from asthma in New Zealand from 1969 to 1976: a case-control study'. *Clinical Pharmacology & Therapeutics* 51 (1992), pp.566–71.

14. W. Bown, 'Warning letter links asthma deaths to drugs'. *New Scientist*, 27 July 1991; N. Pearce, 'Adverse reactions, social responses: a tale of two asthma mortality epidemics'. In P. Davis (ed), *Contested ground: public purpose and private interest in the regulation of prescription drugs.* New York: Oxford University Press, 1996, pp.57–75.

15. W.O. Spitzer, S. Suissa, P. Ernst, R.I. Horwitz, et al., 'Saskatchewan asthma epidemiology study: results of the nested case-control phase assessing the association between beta-agonists and asthma death and near-death'. Montreal: McGill University, 1991.

16. N. Pearce and M.J. Hensley, 'Epidemiologic studies of beta agonists and asthma deaths'. *Epidemiologic Reviews* 20 (1998), pp.173–86.

17. 'The β–agonist dilemma'. *New England Journal of Medicine* 326 (1992), pp.560–1.

18. M.J. Hensley, 'Fenoterol and death from asthma'. *Medical Journal of Australia* 156 (1992), p.882.

19. J.B. McKinlay to N Pearce, 30 March 1992. Copy in author's possession.

20. 'Asthma drug use to be restricted'. *Evening Post*, Wellington, 21 December 1989; *Australian Doctor*, 16 March 1989; N. Pearce, R. Beasley and J. Crane. Mortality of bronchial asthma. In *5th West Pacific Allergy Symposium & 7th Korea–Japan Joint Allergy Symposium.* Seoul, Korea: Monduzzi Editore, 1997.

21. 'beta–agonists and asthma'. *Lancet* 339 (1992), p.422.

22. P.D. Stolley. 'Transcript of proceedings. Pulmonary-allergy drugs Advisory Committee meeting'. Washington, DC: Miller Reporting Company, 1991.

23. N. Pearce, R. Beasley, J. Crane, C. Burgess, et al., 'End of the New Zealand asthma mortality epidemic'. *Lancet* 345 (1995), pp.41–44.

CONCLUSION

1. M. Planck, *Scientific autobiography and other papers.* New York: Greenwood Press, 1949; K.J. Rothman, 'The rise and fall of epidemiology, 1950–2000 AD'. *New England Journal of Medicine* 304 (1981), pp.600–2.

2. R. Doll and A. Bradford Hill, 'Smoking and carcinoma of the lung: preliminary report'. *British Medical Journal* 2 (1950), pp.739–48; E.L. Wynder and E.A. Graham, 'Tobacco smoking as a possible etiologic factor in bronchiogenic carcinoma: a study of 684 proved cases'. *Journal of the American Medical Association* 143 (1950), pp.329–36.

3. R. Kluger, *Ashes to ashes: America's hundred-year cigarette war, the public health, and the unabashed triumph of Philip Morris.* New York: Vintage Books, 1997.

4. 'Asthma drug controversy climaxes with government decision to restrict use'. *Epidemiology Monitor* 11(3) (1990), pp.1–5.

5. P.D. Stolley, interview with Al Morrison, Wellington, 30 June 1989.

6. P. Meier, 'Damned lies and expert witnesses'. *Journal of the American Statistical Association* 394 (1986), pp.269–76.

7. D.A. Savitz, S. Greenland, P.D. Stolley and J.L. Kelsey, 'Scientific standards of criticism: a reaction to "Scientific standards in epidemiologic studies of the menace of daily life." by A.R. Feinstein'. *Epidemiology* 1 (1990), pp.78–83.

8. S. Greenland, 'Science versus advocacy: the challenge of Dr Feinstein'. *Epidemiology* 1 (1990), pp.64–72.

9. P.D. Stolley, 'A public health perspective from academia'. In B.L. Strom (ed), *Pharmacoepidemiology.* New York: Churchill Livingstone, 1989.

10. *Epidemiology Monitor*, 1990, p.4.
11. S.R. Cartwright, 'The report of the Committee of Inquiry into Allegations Concerning the Treatment of Cervical Cancer at National Women's Hospital and into Other Related Matters'. Auckland: Government Printing Office, 1988.
12. C. Paul, 'The New Zealand cervical-cancer study – could it happen again'. *British Medical Journal* 297 (1988), pp.533–9.
13. C. Paul, 'Internal and external morality in medicine: lessons from New Zealand'. *British Medical Journal* 320 (2000), pp.490–503.
14. W.A. McIndoe, M.R. McLean, R.W. Jones and P.R. Mullen, 'The invasive potential of carcinoma in situ of the cervix'. *Obstetrics and Gynaecology* 64 (1984), pp.451–8.
15. S. Coney, *The unfortunate experiment*. Auckland: Penguin Books, 1988; S. Coney and P. Bunkle, 'An "unfortunate experiment" at National Women's'. *Metro*, June 1987.
16. A.P. Duffy, D.K. Barrett and M.A. Duggan, 'Report of the Ministerial Inquiry into the Under-reporting of Cervical Smear Abnormalities in the Gisborne Region'. Wellington: Cervical Screening Inquiry, 2001.
17. J.D. Watson, *The double helix: a personal account of the discovery of the structure of DNA*. New York: Simon & Schuster, 1968; M. Wilkins, *The third man of the double helix: the autobiography of Maurice Wilkins*. Oxford: Oxford University Press, 2003.
18. J. Law, *Big pharma: how the world's biggest drug companies control illness*. London: Constable & Robinson, 2006.
19. I. Kawachi and J. Lexchin, 'Doctors and the drug industry: therapeutic information or pharmaceutical promotion?' In P.B. Davis (ed), *For health or profit: the pharmaceutical industry in New Zealand*. Auckland: Oxford University Press, 1992.
20. N.M. Kaplan, 'The support of continuing medical education by pharmaceutical companies'. *New England Journal of Medicine* 300 (1979), pp.194–6.
21. M.M. Chren, C.S. Landefeld and T.H. Murray, 'Doctors, drug companies and gifts'. *Journal of the American Medical Association* 262 (1989), pp.3448–51.
22. R. Reid, *Microbes and men*. London: BBC, 1974.
23. A. Cochrane, Horizon: The trouble with medicine. BBC TV, 1976.
24. J.M. Elwood and D.C.G. Skegg, 'Review of studies relating to prescribed fenoterol and death from asthma in New Zealand'. Wellington: Department of Health, 1989; J.M. Elwood, 'Prescribed fenoterol and deaths from asthma in New Zealand – second report'. Dunedin: University of Otago, 1989.
25. S. Coney, 'A record of pussyfooting'. *Sunday Times*, Wellington, 7 May 1989.
26. H. Clark, 'Pharmaceutical costs and regulation: from the minister's desk'. In P.B. Davis (ed), *For health or profit: the pharmaceutical industry in New Zealand*. Auckland: Oxford University Press, 1992.
27. M.R. Sears, H.H. Rea, R. Beaglehole, A.J.D. Gillies, et al., 'Asthma mortality in New Zealand: a two year national study'. *New Zealand Medical Journal* 98 (1985), pp.271–5.
28. P.D. Stolley, 'A public health perspective from academia'. In B.L. Strom (ed), *Pharmacoepidemiology*. New York: Churchill Livingstone, 1989, p.54.

Bibliography

BOOKS AND BOOK CHAPTERS

Beasley, R. and N. Pearce (eds). *The role of beta receptor agonist therapy in asthma mortality*. New York: CRC Press, 1993.

Berra, Y., *The Yogi Book*. New York, NY: Workman Press, 1998.

Brodeur, P., *Outrageous misconduct: the asbestos industry on trial*. New York: Pantheon, 1985.

Clark, H. 'Pharmaceutical costs and regulation: from the minister's desk'. In P.B. Davis (ed), *For health or profit: the pharmaceutical industry in New Zealand*. Auckland: Oxford University Press, 1992.

Coney, S., *The unfortunate experiment*. Auckland: Penguin Books, 1988.

Kawachi, I. and J. Lexchin. 'Doctors and the drug industry: therapeutic information or pharmaceutical promotion?' In P. B. Davis (ed), *For health or profit: the pharmaceutical industry in New Zealand*. Auckland: Oxford University Press, 1992.

Kluger, R., *Ashes to ashes: America's hundred-year cigarette war, the public health, and the unabashed triumph of Philip Morris*. New York: Vintage Books, 1997.

Law, J., *Big pharma: how the world's biggest drug companies control illness*. London: Constable & Robinson, 2006.

London Sunday Times Insight Team, *Suffer the children: the story of thalidomide*. New York: Viking Press, 1979.

Matsui, T. 'Asthma death and β–2 agonists.' In K. Shinomiya (ed), *Current advances in pediatric allergy and clinical immunology. Selected proceedings from the 32nd Annual Meeting of the Japanese Society of Pediatric Allergy and Clinical Immunology. Tokyo, Japan*. Tokyo: Churchill Livingstone, 1996, pp.161–4.

Neuberger, M.B., *Smoke screen: Tobacco and the public welfare*. Englewood Cliffs: Prentice-Hall, 1963.

Pearce, N. 'Adverse reactions: the fenoterol saga'. In P. Davis (ed), *For health or profit: the pharmaceutical industry in New Zealand*. Auckland: Oxford University Press, 1992, pp.75–97.

Pearce, N. 'Adverse reactions, social responses: a tale of two asthma mortality epidemics'. In P. Davis (ed), *Contested ground: public purpose and private interest in the regulation of prescription drugs*. New York: Oxford University Press, 1996, pp.57–75.

Pearce, N. 'Public health and the precautionary principle'. In, *The precautionary principle, public health, protection of children and sustainability*. Rome: WHO Regional Office for Europe, 2004, pp.49–62.

Pearce, N., R. Beasley, C. Burgess and J. Crane, *Asthma epidemiology: principles and methods*. New York: Oxford University Press, 1998.

Pearce, N., R. Beasley, J. Crane and C. Burgess. 'Epidemiology of asthma mortality'. In S.T. Holgate and W.W. Busse (eds), *Asthma and rhinitis*. Oxford: Blackwell Scientific, 2001, pp.56–69.

Pearce, N., R. Beasley, J. Crane and C. Burgess. 'Pharmacoepidemiology of asthma deaths'. In H. Tilson (ed), *Pharmacoepidemiology: an introduction*. Cincinnati, OH: Harvey Whitney, 1998, pp.473–94.

Pearce, N., J. Douwes and R. Beasley. 'Asthma'. In R. Detels, J. McEwen, R. Beaglehole and H. Tanaka (eds), *Oxford textbook of public health*. Oxford: Oxford University Press, 2002, pp.1255–77.

Planck, M., *Scientific autobiography and other papers*. New York: Greenwood Press, 1949.

Reid, R., *Microbes and men*. London: BBC, 1974.

Salter, H.H., *On asthma: its pathology and treatment*. London, 1860.

Santayana, G., *Life of reason: reason in common sense*. New York: Scribner's, 1905.

Stolley, P.D. 'A public health perspective from academia'. In B. L. Strom (ed), *Pharmacoepidemiology*. New York: Churchill Livingstone, 1989.

Taubes, G., *Bad science: the short life and weird times of cold fusion*. New York: Random House, 1993.

Tomatis, R., *Il Fuoriscito*. Milan: Sirino, 2005.

Watson, J.D., *The double helix: a personal account of the discovery of the structure of DNA*. New York: Simon & Schuster, 1968.

Wilkins, M., *The third man of the double helix: the autobiography of Maurice Wilkins*. Oxford: Oxford University Press, 2003.

ACADEMIC PAPERS

'Asthma deaths: a question answered'. *British Medical Journal* 1972/4, pp.443–4.

'Fatal asthma'. *Lancet* 1979/2, pp.337–8.

'Death from asthma in New Zealand'. *New Zealand Medical Journal* 95 (1982), pp.692–3.

'Asthma drug controversy climaxes with government decision to restrict use'. *Epidemiology Monitor* 11(3) (1990), pp.1–5.

'β-2 agonists in asthma: relief, prevention, morbidity'. *Lancet* 336 (1990), pp.1391–6.

'The β-agonist dilemma'. *New England Journal of Medicine* 326 (1992), pp.560–1.

'β-agonists and asthma'. *Lancet* 339 (1992), p.422.

Anto, J.M. and J. Sunyer. 'Epidemiologic Studies of Asthma Epidemics in Barcelona'. *Chest* 98 (1990), pp.S185–S190.

Beaglehole, R., R. Jackson, M. Sears and H.H. Rea. 'Asthma mortality in New Zealand: a review with some policy implications'. *New Zealand Medical Journal* 100 (1987), pp.231–4.

Beasley, R., C. Burgess, N. Pearce, K. Woodman, et al. 'Confounding by severity does not explain the association between fenoterol and asthma death'. *Clinical & Experimental Allergy.* 24 (1994), pp.660–668.

Beasley, R., J. Crane, C. Burgess and N. Pearce. 'Bronchial responsiveness during regular fenoterol therapy'. *New Zealand Medical Journal.* 104 (1991), pp.124.

Beasley, R., J. Crane, C. Burgess, N. Pearce, et al. 'Fenoterol and severe asthma mortality'. *New Zealand Medical Journal* 102 (1989), pp.294–5.

Beasley, R., S. Nishima, N. Pearce and J. Crane. 'Beta-agonist therapy and asthma mortality in Japan'. *Lancet.* 351 (1998), pp.1406–7.

Beasley, R., N. Pearce, J. Crane, H. Windom, et al. 'Asthma mortality and inhaled beta agonist therapy'. *Australian & New Zealand Journal of Medicine* 21 (1991), pp.753–63.

Benatar, S.R. 'Fatal asthma'. *New England Journal of Medicine* 314 (1986), pp.423–9.

Bremner, P., C. Burgess, R. Beasley, K. Woodman, et al. 'Nebulized fenoterol causes greater cardiovascular and hypokalaemic effects than equivalent bronchodilator doses of salbutamol in asthmatics'. *Respiratory Medicine* 86 (1992), pp.419–23.

Bremner, P., C.D. Burgess, J. Crane, D. McHaffie, et al. 'Cardiovascular effects of fenoterol under conditions of hypoxaemia'. *Thorax* 47 (1992), pp.814–7.

Bremner, P., R. Siebers, J. Crane, R. Beasley, et al. 'Partial vs full β-receptor agonism, a clinical study of inhaled albuterol and fenoterol'. *Chest* 109 (1996), pp.957–62.

Buist, A.S., P.G.J. Burney, A.R. Feinstein, R.I. Horwitz, et al. 'Fenoterol and fatal asthma'. *Lancet* 1989/1, p.1071.

Campbell, A.H. 'Mortality from asthma and bronchodilator aerosols'. *Medical Journal of Australia* 1976/1, pp.386–91.

Chren, M.M., C.S. Landefeld and T.H. Murray. 'Doctors, drug companies and gifts'. *Journal of the American Medical Association* 262 (1989), pp.3448–51.

Cohen, E. and A. Van As. 'A haemodynamic study on fenoterol (Berotec) in the baboon'. *Medical Proceedings* (1972), pp.24–27.

Collins, J.M., D.G. McDevitt, R.G. Shanks and J.G. Swanton. 'The cardiotoxicity of isoprenaline during hypoxia'. *British Journal of Pharmacology* 36 (1969), pp.35–45.

Copeland, K.T., H. Checkoway, A.J. McMichael and R.H. Holbrook. 'Bias due to misclassification in the estimation of relative risk'. *American Journal of Epidemiology* 105 (1977), pp.488–95.

Crane, J., C. Burgess and R. Beasley. 'Cardiovascular and hypokalaemic effects of inhaled salbutamol, fenoterol and isoprenaline'. *Thorax* 44 (1989), pp.136–40.

Crane, J., N. Pearce, C. Burgess and R. Beasley. 'Fenoterol and fear of flying'. *New Zealand Medical Journal*. 102 (1989), p.514.

Crane, J., N. Pearce, C. Burgess, K. Woodman, et al. 'Markers of risk of asthma death or readmission in the 12 months following a hospital admission for asthma'. *International Journal of Epidemiology* 21 (1992), pp.737–44.

Crane, J., N. Pearce, A. Flatt, C. Burgess, et al. 'Prescribed fenoterol and death from asthma in New Zealand, 1981–83: case-control study'. *Lancet* 1989/1, pp.917–22.

Doll, R. and A. Bradford Hill. 'Smoking and carcinoma of the lung: preliminary report'. *British Medical Journal* 1950/2, pp.739–48.

Doll, R. and A. Bradford Hill. 'Mortality in relation to smoking: ten years' observations of British doctors'. *British Medical Journal* 1964/1, pp.1399–410.

Ellul-Micallef, R. 'Asthma: a look at the past'. *British Journal of Diseases of the Chest* 70 (1976), pp.112–16.

Elwood, J.M. 'Fenoterol: the evidence leading to restriction of its use'. *New Zealand Medical Journal* 103 (1990), pp.395–7.

Esdaile, J.M., A.R. Feinstein and R.I. Horwitz. 'A reappraisal of the United Kingdom epidemic of fatal asthma'. *Archives of Internal Medicine* 147 (1987), pp.543–9.

Fraser, P.M., F.E. Speizer, S.D.M. Waters, R. Doll, et al. 'The circumstances preceding death from asthma in young people in 1968 to 1969'. *British Journal of Diseases of the Chest* 65 (1971), pp.71–84.

Gandevia, B. 'Pressurized sympathomimetic aerosols and their lack of relationship to asthma mortality in Australia'. *Medical Journal of Australia* i (1973), pp.273–7.

Grainger, J., K. Woodman, N. Pearce, J. Crane, et al. 'Prescribed fenoterol and death from asthma in New Zealand, 1981–7: a further case-control study'. *Thorax* 46 (1991), pp.105–11.

Grant, I.W.B. 'Asthma in New Zealand'. *British Medical Journal* 286 (1983), pp.374–7.

Grant, I.W.B. 'Asthma in New-Zealand – Reply'. *British Medical Journal* 286 (1983), p.718.

Grant, I.W.B. 'Fenoterol and asthma deaths in New Zealand'. *New Zealand Medical Journal* 103 (1990), pp.160–1.

Greenberg, M.J. and A. Pines. 'Pressurized aerosols in asthma'. *British Medical Journal* 1 (1967), p.563.

Greenland, S. 'Science versus advocacy: the challenge of Dr Feinstein'. *Epidemiology* 1 (1990), pp.64–72.

Hendeles, L. and M. Weinberger. 'Nonprescription sale of inhaled metaproterenol – déjà vu'. *New England Journal of Medicine*. 310 (1984), pp.207–8.

Hensley, M.J. 'Fenoterol and death from asthma'. *Medical Journal of Australia* 156 (1992), p.882.

Inman, W.H.W. and A.M. Adelstein. 'Rise and fall of asthma mortality in England and Wales in relation to use of pressurized aerosols'. *Lancet* 1969/2, pp.279–85.

Jackson, R.T., R. Beaglehole, H.H. Rea and D.C. Sutherland. 'Mortality from asthma: a new epidemic in New Zealand'. *British Medical Journal* 285 (1982), pp.771–4.

Jenkins, M.A., S.F. Hurley, G. Bowes and J.J. McNeil. 'Use of antiasthmatic drugs in Australia'. *Medical Journal of Australia*. 153 (1990), pp.32–38.

Keating, G., E.A. Mitchell, R. Jackson, R. Beaglehole, et al. 'Trends in sales of drugs for asthma in New Zealand, Australia and the United Kingdom, 1975–81'. *British Medical Journal* 289 (1984), pp.348–51.

Keeney, E.L. 'The history of asthma from Hippocrates to Meltzer'. *Journal of Allergy & Clinical Immunology* 35 (1964), pp.215–26.

Kundig, H. 'Preliminary pharmacological and toxicological studies in the baboon (Papio ursinus) on a new β2 adrenegeric stimulant, fenoterol (Berotec)'. *Medical Proceedings* (1972), pp.9–14.

Lanes, S.F. and A.M. Walker. 'Do pressurized bronchodilator aerosols cause death among asthmatics?' *American Journal of Epidemiology* 125 (1987), pp.755–60.

Last, J. 'Epidemiology and ethics'. *Lancet* 336 (1990), p.497.

Mackay, R.J. and J.H. van der Westhuyzen. 'Asthma in New Zealand: some answers, more questions'. *New Zealand Medical Journal*. 101 (1988), pp.835–6.

McIndoe, W.A., M.R. McLean, R.W. Jones and P.R. Mullen. 'The invasive potential of carcinoma in situ of the cervix'. *Obstetrics and Gynaecology* 64 (1984), pp.451–8.

Meier, P. 'Damned lies and expert witnesses'. *Journal of the American Statistical Association* 394 (1986), pp.269–76.

Munro Ford, R. 'Asthma and aerosol sprays'. *Medical Journal of Australia* 1966/4, pp.667–8.

O'Donnell, T.V. 'Asthma in New Zealand: a response'. *New Zealand Medical Journal* 96 (1983), pp.163–5.

O'Donnell, T.V., H.H. Rea, P.E. Holst and M.R. Sears. 'Fenoterol and fatal asthma'. *Lancet* 1989/1, pp.1070–1.

Olson, L.G. 'Acute severe asthma: what to do until the ambulance arrives'. *New Ethicals* 25 (1988), pp.105–16.

Patrick, P.R. and J.I. Tonge. 'Asthma and aerosol sprays'. *Medical Journal of Australia* 1966/2, p.668.

Paul, C. 'The New Zealand cervical-cancer study – could it happen again'. *British Medical Journal* 297 (1988), pp.533–9.

Paul, C. 'Internal and external morality in medicine: lessons from New Zealand'. *British Medical Journal* 320 (2000), pp.490–503.

Pearce, N., R. Beasley and J. Crane. Mortality of bronchial asthma. In *5th West Pacific Allergy Symposium & 7th Korea–Japan Joint Allergy Symposium*. Seoul, Korea: Monduzzi Editore, 1997.

Pearce, N., R. Beasley, J. Crane, C. Burgess, et al. 'End of the New Zealand asthma mortality epidemic'. *Lancet* 345 (1995), pp.41–44.

Pearce, N., J. Crane, C. Burgess, R. Jackson, et al. 'Beta agonists and asthma mortality: déjà vu'. *Clinical & Experimental Allergy* 21 (1991), pp.401–10.

Pearce, N., J. Grainger, M. Atkinson, J. Crane, et al. 'Case-control study of prescribed fenoterol and death from asthma in New Zealand, 1977–81'. *Thorax* 45 (1990), pp.170–5.

Pearce, N. and M.J. Hensley. 'Epidemiologic studies of beta agonists and asthma deaths'. *Epidemiologic Reviews* 20 (1998), pp.173–86.

Poole, C., S.F. Lanes and A.M. Walker. 'Fenoterol and fatal asthma'. *Lancet* 335 (1990), p.920.

Randall, T. 'Ethics of receiving gifts considered.' *New England Journal of Medicine*. 265 (1991), pp.442–3.

Rea, H.H., R. Scragg, R. Jackson, R. Beaglehole, et al. 'A case-control study of deaths from asthma'. *Thorax* 41 (1986), pp.833–9.

Read, J. 'Pressurized aerosols in asthma'. *British Medical Journal* 1967/1, p.699.

Read, J. 'The reported increase in mortality from asthma: a clinico-functional analysis'. *Medical Journal of Australia* 1968/1, pp.879–91.

Rothman, K.J. 'The rise and fall of epidemiology, 1950–2000 AD'. *New England Journal of Medicine* 304 (1981), pp.600–2.

Sackett, D.L., H.S. Shannon and G.W. Browman. 'Fenoterol and Fatal Asthma'. *Lancet* 335 (1990), pp.45–46.

Savitz, D.A., S. Greenland, P.D. Stolley and J.L. Kelsey. 'Scientific standards of criticism: a reaction to "Scientific standards in epidemiologic studies of the menace of daily life." by A.R. Feinstein'. *Epidemiology* 1 (1990), pp.78–83.

Sears, M.R. 'Asthma in New Zealand: some answers, more questions'. *New Zealand Medical Journal*. 101 (1988), pp.598–600.

Sears, M.R. and R. Beaglehole. 'Asthma morbidity and mortality: New Zealand'. *Journal of Allergy & Clinical Immunology* 80 (1987), pp.383–8.

Sears, M.R., H.H. Rea, R. Beaglehole, A.J.D. Gillies, et al. 'Asthma mortality in New Zealand: a two year national study'. *New Zealand Medical Journal* 98 (1985), pp.271–5.

Sears, M.R., H.H. Rea, G. De Boer, R. Beaglehole, et al. 'Accuracy of certification of deaths due to asthma – a national study'. *American Journal of Epidemiology* 124 (1986), pp.1004–11.

Sears, M.R., H.H. Rea, H. Fenwick, A.J.D. Gillies, et al. '75 deaths in asthmatics prescribed home nebulisers'. *British Medical Journal* 294 (1987), pp.477–80.

Sears, M.R. and D.R. Taylor. 'The β-2 agonist controversy: observations, explanations and relationship to asthma epidemiology'. *Drug Safety* 11 (1994), pp.259–83.

Sears, M.R., D.R. Taylor, C.G. Print, D.C. Lake, et al. 'Regular inhaled beta-agonist treatment in bronchial asthma'. *Lancet.* 336 (1990), pp.1391–6.

Speizer, F.E., R. Doll and P. Heaf. 'Observations on recent increase in mortality from asthma'. *British Medical Journal* 1968/1, pp.335–9.

Speizer, F.E., R. Doll, P. Heaf and L.B. Strang. 'Investigation into use of drugs preceding death from asthma'. *British Medical Journal* 1968/1, pp.339–43.

Spitzer, W.O., S. Suissa, P. Ernst, R.I. Horwitz, et al. 'The use of beta-agonists and the risk of death and near death from asthma'. *New England Journal of Medicine* 326 (1992), pp.501–6.

Stolley, P.D. 'Why the United States was spared an epidemic of deaths due to asthma'. *American Review of Respiratory Disease* 105 (1972), pp.883–90.

Stolley, P.D. 'Asthma deaths and bronchodilator aerosols – reply'. *American Review of Respiratory Disease* 107 (1973), pp.1078–9.

Stolley, P.D. 'When genius errs: Fisher, R.A. and the lung cancer controversy'. *American Journal of Epidemiology* 133 (1991), pp.416–25.

Stolley, P.D. and T. Lasky. 'The bellman always rings thrice'. *Annals of Internal Medicine* 118 (1993), p.158.

Stolley, P.D. and R. Schinnar. 'Association between asthma mortality and isoproterenol aserosols: a review'. *Preventive Medicine* 7 (1978), pp.319–38.

Strunk, R.C., D.A. Mrazek, G.S. Wolfson Fuhrmann and J.F. LaBrecque. 'Physiologic and psychologic characteristics associated with deaths due to asthma in childhood'. *Journal of the American Medical Association* 254 (1985), pp.1193–8.

Suissa, S., P. Ernst, J.F. Boivin, R.I. Horwitz, et al. 'A cohort analysis of excess mortality in asthma and the use of inhaled beta-agonists'. *American Journal of Respiratory and Critical Care Medicine* 149 (1994), pp.604–10.

Sutherland, D.C. and H.H. Rea. 'Asthma in New Zealand'. *British Medical Journal* 286 (1983), pp.717–8.

Tandon, M.K. 'Cardiopulmonary effects of fenoterol and salbutamol aerosols'. *Chest.* 77 (1980), pp.429–31.

Thornley, P.E. and K.P. Dawson. 'Asthma in New Zealand'. *British Medical Journal* 286 (1983), pp.890–1.

Unger, L. and M.C. Harris. 'Stepping stones in allergy'. *Annals of Allergy* 32 (1974), pp.214–30.

van Metre, T.E. 'Adverse effects of inhalation of excessive amounts of nebulized isoproterenol in status asthmaticus'. *Journal of Allergy* 43 (1969), pp.101–13.

Venning, G.R. 'Identification of adverse reactions to new drugs. I. What have been the important adverse reactions since thalidomide?' *British Medical Journal* 286 (1983), pp.199–202.

Wardell, W.M. 'Nonprescription sales of metaproterenol aerosols'. *New England Journal of Medicine* 311 (1984), pp.405–6.

Weinberger, M. and L. Hendeles. 'Nonprescription sales of metaproterenol aerosols'. *New England Journal of Medicine* 311 (1984), p.406.

Wilson, J.D., D.C. Sutherland and A.C. Thomas. 'Has the change to beta-agonists combined with oral theophylline increased cases of fatal asthma?' *Lancet* 1981/1, pp.1235–7.

Windom, H.H., C.D. Burgess, J. Crane, N. Pearce, et al. 'The self-administration of inhaled beta agonist drugs during severe asthma'. *New Zealand Medical Journal* 103 (1990), pp.205–7.

Wong, C.S., I.D. Pavord, J. Williams, J.R. Britton, et al. 'Bronchodilator, cardiovascular, and hypokalemic effects of fenoterol, salbutamol, and terbutaline in asthma'. *Lancet* 336 (1990), pp.1396–9.

Woodman, K., N. Pearce, R. Beasley, C. Burgess, et al. 'Albuterol and deaths from asthma in New Zealand from 1969 to 1976: a case-control study'. *Clinical Pharmacology & Therapeutics* 51 (1992), pp.566–71.

Woolcock, A.J. Aerosol bronchodilator therapy in asthma. In *10th Asia Pacific Congress on*

Diseases of the Chest. Amsterdam: Excerpta Medica, Elsevier, 1985.
Woolcock, A.J., M.P. Alpers and V.M. Hurry. 'An epidemic of asthma in Papua New Guinea'. *Australian and New Zealand Journal of Medicine* 18 (1988), pp.544.
Wynder, E.L. and E.A. Graham. 'Tobacco smoking as a possible etiologic factor in bronchiogenic carcinoma: a study of 684 proved cases'. *Journal of the American Medical Association* 143 (1950), pp.329–36.

NEWSPAPER AND MAGAZINE ARTICLES
Bown, W., 'Warning letter links asthma deaths to drugs'. *New Scientist*, 27 July 1991.
Chisholm, D., 'Asthma drug given to delay labour'. *Sunday Star*, Auckland, 7 May.
Coney, S., 'A record of pussyfooting'. *Sunday Times*, Wellington, 7 May 1989.
Coney, S. and P. Bunkle, 'An "unfortunate experiment" at National Women's'. *Metro*, June 1987.
Cowie, C., 'Asthma drug maker under pressure'. *Evening Post*, Wellington, 2 May 1989.
Cowie, C., 'Public health group seeks asthma inquiry'. *Evening Post*, Wellington, 8 May 1989.
Cowie, C., 'Quick asthma study back-up unlikely'. *Evening Post*, Wellington, 1 May 1989.
Dawson, D., 'Physicians back asthma findings'. *Sunday Times*, Wellington, 30 April 1989.
Drent, R., 'Asthma drug blamed again'. *Dominion*, Wellington, 29 June 1989.
Drent, R., 'Asthma study unsure – council'. *Dominion*, Wellington, 26 May 1989.
Drent, R., 'Harm to doctors' training feared'. *Dominion*, Wellington, 12 July 1989.
Horswell, I., 'Ape deaths link with asthma drug'. *Sunday Star*, Auckland, 4 May 1989.
Morrison, A., 'Controversy rages over asthma drug'. *Sunday Times*, Wellington, 2 July 1989.
Morrison, A., 'Council quits research funded by drug firm'. *Dominion.*, Wellington, 17 October 1989.
Morrison, A., 'Power plays in the fight to save a suspect drug'. *Dominion*, Wellington, 22 September 1989.
Nixon, J., 'Association in strife as drug kings wage war.' *National Business Review*, Auckland, 22 May 1989.
Quayle, S., 'Warning on asthma drug'. *Australian Doctor*, 16 March 1989.
Ray, G., 'Researchers say asthma puffer drugs need rethink'. *Newcastle Herald*, Newcastle, Australia, 29 June 1989.
Wilkins, I., 'Asthmatics in dark as drug study widens research row'. *Sunday Star*, Auckland, 7 May 1989.

REPORTS
Cartwright, S.R., 'The report of the Committee of Inquiry into Allegations Concerning the Treatment of Cervical Cancer at National Women's Hospital and into Other Related Matters'. Auckland: Government Printing Office, 1988.
Duffy, A.P., D.K. Barrett and M.A. Duggan, 'Report of the Ministerial Inquiry into the Under-reporting of Cervical Smear Abnormalities in the Gisborne Region'. Wellington: Cervical Screening Inquiry, 2001.
Elwood, J.M., 'Prescribed fenoterol and deaths from asthma in New Zealand – second report'. Dunedin: University of Otago, 1989.
Elwood, J.M. and D.C.G. Skegg, 'Review of studies relating to prescribed fenoterol and death from asthma in New Zealand'. Wellington: Department of Health, 1989.
Moore, A., M. Ardagh, D. Bramley, A. Bray, et al., 'Review of the current processes for ethical review of health and disability research in New Zealand: report to the Minister of Health'. Wellington: National Ethics Advisory Committee, 2004.

Index

Van der Westhuyzen, J.H., 44
Ventolin, *see* salbutamol

Wahl, J., 96
Walker, A., 27, 98, 101, 139, 153
Wardell, W., 26, 96
Waters, M., 99
Watson, J.D., 83, 183
Wellington Hospital, 12, 60, 70, 109
Wellington Medical School, 1, 48, 50, 52,
    55, 58, 60, 61, 62, 75, 83, 88, 103, 122,
    123

Wellington School of Medicine, *see*
    Wellington Medical School
West Germany, *see* Germany
White P., 89
Wilkins, M., 183
Wilson, D., 32–34, 41, 44, 45–47, 51, 57,
    73–74, 76, 87–88, 89, 94, 104, 106, 114,
    118–9, 129–30, 152, 172
Woolcock, A., 87, 107, 121, 151–2
World Health Organisation, 59, 64, 181

Yugoslavia, 158